26 Advances in Polymer Science

Fortschritte der Hochpolymeren-Forschung

Edited by H.-J. CANTOW, Freiburg i. Br. · G. DALL'ASTA, Cesano Maderno
K. DUŠEK, Prague · J. D. FERRY, Madison · H. FUJITA, Osaka
M. GORDON, Colchester · W. KERN, Mainz · G. NATTA, Milano
S. OKAMURA, Kyoto · C. G. OVERBERGER, Ann Arbor · T. SAEGUSA, Kyoto
G. V. SCHULZ, Mainz · W. P. SLICHTER, Murray Hill · J. K. STILLE, Fort Collins

With 61 Figures

Springer-Verlag
Berlin Heidelberg GmbH 1978

Editors

Prof. Dr. HANS-JOACHIM CANTOW, Institut für Makromolekulare Chemie der Universität, Stefan-Meier-Str. 31, 7800 Freiburg i. Br., BRD

Prof. Dr. GINO DALL'ASTA, SNIA VISCOSA – Centro Sperimentale, Cesano Maderno (MI), Italia

Prof. Dr. KAREL DUŠEK, Institute of Macromolecular Chemistry, Czechoslovak Academy of Sciences, 162 06 Prague 616, ČSSR

Prof. Dr. JOHN D. FERRY, Department of Chemistry, The University of Wisconsin, Madison 6, Wisconsin 53706, U.S.A.

Prof. Dr. HIROSHI FUJITA, Osaka University, Department of Polymer Science, Toyonaka, Osaka, Japan

Prof. Dr. MANFRED GORDON, University of Essex, Department of Chemistry, Wivenhoe Park, Colchester C04 3 SQ, England

Prof. Dr. WERNER KERN, Institut für Organische Chemie der Universität, 6500 Mainz, BRD

Prof. Dr. GIULIO NATTA, Istituto di Chimica Industriale del Politecnico, Milano, Italia

Prof. Dr. SEIZO OKAMURA, Department of Polymer Chemistry, Kyoto University, Kyoto, Japan

Prof. Dr. CHARLES G. OVERBERGER, The University of Michigan, Department of Chemistry, Ann Arbor, Michigan 48 104, U.S.A.

Prof. TAKEO SAEGUSA, Kyoto University, Department of Synthetic Chemistry, Faculty of Engineering, Kyoto, Japan

Prof. Dr. GÜNTER VICTOR SCHULZ, Institut für Physikalische Chemie der Universität, 6500 Mainz, BRD

Dr. WILLIAM P. SLICHTER, Bell Telephone Laboratories Incorporated, Chemical Physics Research Department, Murray Hill, New Jersey 07 971, U.S.A.

Prof. Dr. JOHN K. STILLE, Colorado State University, Department of Chemistry, Fort Collins, CO 805 23, U.S.A.

ISBN 978-3-662-15450-2   ISBN 978-3-540-35896-1 (eBook)
DOI 10.1007/978-3-540-35896-1

Library of Congress Catalog Card Number 61-642

© by Springer-Verlag Berlin Heidelberg 1978
Originally published by Springer-Verlag Berlin Heidelberg New York in 1978
Softcover reprint of the hardcover 1st edition 1978

2152/3140 – 543210

Contents

Molecular Mobility, Deformation and Relaxation Processes in Polymers

Werner Holzmüller

Sektion Physik, Karl-Marx-Universität, 7010 Leipzig, GDR

Table of Contents

List of Symbols

A_{el} or A	Elastic potential caused by an external stress
A_{diel}	Electric potential caused by an external electric field
A, B, C_1, C_2, C_3, C_4	Constants
a	Characteristic quasiviscosity in the WLF equation
c	Concentration
D	Diffusion constant
E	Electric field strength (E_{or} orientation field strength)
E or E_m	Young modules (E_m extended to molecular dimension in our quasicubic model)
F	Force
G or G', G''	Shear modules (G' real, G'' imaginary component) (G_0 for purely elastic deformation)
f_r	Reducing factor
h	Planck constant ($h = 6{,}62 \ 10^{-34}$ Joule sec)
I	Diffusion current
K	Modules of compressibility
k	Boltzmann constant ($k = 1{,}380 \ 10^{-23}$ Joule/°K)
L	Length of a capillary
L	Laplace transformation
M	Torque
m, n	Constants characterizing the amorphous state
m	Mass of a flowing unit
n	Number of coupled processes, characterizing a network $N \approx 1/n$
p	Pressure
P	Electric polarisation (P_{def} deformation, P_{or} orientation polarisation)
p_x	Momentum coordinate
r	Spatial coordinate
r_0	Radius of a flowing unit, distance between the maximum and the saddle point in the Lennard-Jones potential
R	Radius of a capillary or cylinder (R_i inner, R_a outer radius)
S	Entropy
T	Temperature (T_g glass transition temperature)
z	Number of molecular dislocations in the direction of an external stress per unit area
z_0	Number of flowing units or segments (per unit area)
U	Potential energy (U_i, U_k potential energy at a minimum of an energy hyperplane, U_{ov} energy needed to overcome an energy barrier in special cases, ΔU activation energy)
v	Velocity
V	Volume
W	Probability (W_p probability for molecular dislocation processes)
y, z	Spatial coordinates
α	Partion of flowing units participating in molecular fluctuation processes (molecules have more than one possible conformation)
$\alpha, \beta, \gamma, \delta, \varphi, \psi, \theta$	Special angles
β_{th}	Linear thermal expansion coefficient
γ	Shear rate
ϵ	Elongation $\Lambda = 1 + \epsilon = \dfrac{1 + \Delta l}{1}$
ϵ	Dielectric constant ($\epsilon_0 = 8{,}854 \ 10^{-12}$ As/Vm)
Δ, δ	Differences

χ	Quantity describing an amorphous state
η	Viscosity
μ	Dipole moment
ν_{th}	Average frequency for thermal vibrations, $\nu_{th} = kT/h \approx 10^{13}$
α	Molecular polarisability
τ	Relaxation time; τ_{stress} for stress, τ_{strain} for strain
τ_{fl}	Fluctuation time
ω	Angular velocity, frequency

1. Introduction

Different Possibilities to Understand Deformation and Relaxation

The viscoelastic behavior, flow processes, dielectric losses and all time dependent reactions have been investigated in various ways such as:

1. Hydrodynamical conceptions transferred to molecular dimensions by considering the molecules as billiard balls or ellipsoids in a structureless environments. (Stuart, Saito, Debye, Bueche[1].)

2. Phenomenological considerations, leading to linear differential equations based on simple models *e.g.* Maxwell, Kelvin (Linear theory).

3. Theories assuming on the formation of a free volume and mobility of the molecules within the holes (Williams, Landel, Ferry, Tobolsky, Kovacs and others)[2].

4. Statistical molecular considerations. The knowledge of structure and molecular fields of force allows calculation of the probability of molecular and segmental displacements. These are stimulated by thermal motion and are connected with the existence of molecular holes. The foundations of the theory of rate processes were given by Prandtl[3] and Eyring[4]. Using the Arrhenius formula for the probability of molecular rate processes in the direction and opposite to an external force, Eyring calculated the difference of displacements in both these directions in good agreement with Newtonian and non-Newtonian flow. An exponential dependence on temperature for the viscosity was also found. Under the influence of an external mechanical or electrical force the thermodynamical equilibrium will be disturbed, causing thermal collisions that gradually lead to the formation of a new equilibrium state. The relaxation time is the time necessary for the establishment of $1/e$ of this new equilibrium. The relaxation time depends on the height of the potential barriers or activation energies, which must be overcome.

Besides the original works of Prandtl, Eyring and Holzmüller[5] we find only few attempts to describe the flow processes in terms of realistic chain molecules. They are mainly connected with the dielectric behavior caused by the simplicity of discribing rotation processes leading to an orientation polarisation. Mention should be also given the theories of Kirkwood[6], Fuoss[7], Fröhlich[33], Hoffmann[38], Ishida[35] and Gottlib[39] in this context. These theories can be summarized using the name "barrier theories". The heights of the barriers are not equal, but possess a distribution of the activation energies which results in a spectrum of relaxation processes.

The rate processes being considered in this work are in stronger connection with mechanical deformation. They include the role of the free volume, the superposing of thermal vibrations, the intermolecular chemical binding forces and the influence of external shearing stresses and fields. Moreover, we intend to explain viscous and viscoelastic flow as molecular dislocations and to combine pure elastic deformation with reversible and irreversible flow processes. We wish to describe a generally comprehensive theory connecting the structural conceptions of the chemists to the technical observations in industry. This is only possible with some simplifications: Quasi-cubic structure of the flowing units consisting of molecular segments, spherical atomic fields of force and a single average activation energy in most cases which yields but one average relaxation process.

Molecular dislocation is understood here to mean the change of possible conformations caused by thermal vibrations. Let us consider thermal vibrations as independent oszillations of real particles. The superposition of oszillations with different energies leads in some cases to vibration energies $\geqq \Delta U$, thereby allowing the formation of another conformation. The change of molecular conformations is characterized by the overcoming of a molecular barrier possessing the activation energy ΔU. We sometimes speak of "flip-flop reactions". (In our recent publications we used the expression "phonon" to describe the classical harmonic thermal oszillations. Since the phonons are connected with the entire thermodynamical system and their identification with single particles can lead to misunderstanding, we avoid the expression phonon in this work.)

We are interested only in thermal vibrations from the equilibrium point (minimum) in the direction of a saddle point of the energy hyperplane. There is no friction in the hydrodynamical sense. However a change of the kinetic and the potential energy of the thermal vibrations does take place, due to the fact that in all relaxation and flow processes the kinetic energy of all oszillating systems combined will increase, this the effect of external stress, external electric or magnetic fields. In all cases we observe an excess of vibration energy in relaxation processes.

Some critical remarks must be made:

1. The molecular binding energies obtained from chemical conceptions are lower than the activation energies gained from the slope in the logarithmic representation of the Arrhenius formula.

2. The flowing units for molecular displacements are larger, than the well known sizes of segments and branches in macromolecules.

3. The Arrhenius equation is described by a straight line in the log η, $(1/T)$ diagram. We on the other hand find crooked lines in this representation, implying a smaller decrease of viscosity at the high temperature range as the result of heating. In some publications we gave references which suggested methods for the removal of these deviations:

1. Superposition of more than two thermal vibrations causing molecular dislocations removes the disproportions between the chemical binding energy U_{chem} and the physical activation energy. The Arrhenius formula describes only the process of two thermal vibrations[8].

2. It is impossible to consider independent thermal vibrations connected with single particles at the low temperature range. In solid systems neighboring particles will partially vibrate in phase. Only that part of the thermal vibration energy which is caused by phase differences between neighboring particles stimulates molecular dislocation. At the low temperature range overcoming the energy barrier, will be a rare case and it cannot be calculated using the Arrhenius formula[9].

3. The forming of free volume, increases the fraction α of particles participating in the fluctuation processes between several possible conformations. Intermolecular mobility for amorphous materials begins at the glass transition temperature characterized by an increase of the free volume[10].

If only a few possible conformations exist, coupled dislocation processes occur. These change to single independent rate processes with increasing temperature. Finally all particles take part in molecular dislocation processes. There must be a saturation effect for the part α of mobile molecules. This effect is described by the Williams-Landel-Ferry relation[2, 11]. The combination of statistical dislocations created by the interactions between thermal vibrations and lattice conformations as well as an increase of the free volume describes the viscoelastic behavior completely.

4. In most cases some adherent places between different macromolecules or different segments must be overcome simultaneously. This leads to a diameter of about 10^{-6} cm and to activation energies of 10 to 15 kcal/gmol for the flowing units[10, 11].

5. We find differences between the relaxation of stress and strain[12].

6. In some cases the external shearing energy, that is an elastical potential A, may be compared to the thermal vibration energy. With increasing elastical potential the relaxation time decreases and we find a typical Eyring flow. Finally, a large external stress creates exponential flow and temperature dependent fragmentation.

7. Independent molecular displacements lead simultaneously or consecutively to a system of coupled differential equations characterized by longer relaxation times or by several different relaxation processes.

8. The existence of different barriers ΔU_i calculated by empirical formulas or by quantum mechanics yields a spectrum of relaxation processes (α, β, γ relaxation). Hoffman uses an arbitrary rectangular distribution of relaxation processes[38]. Other authors use a Gaussian distribution for the relaxation processes[14].

9. In contrary to Eyring[4], we use the activation energy ΔU given by chemical binding forces to calculate the probability for barrier processes. Eyring uses the free energy $F = U - TS$ and finds the factor $\exp - S/k$ instead of our temperature dependent coefficient α (Chapter 4). Our considerations are derived from single statistic assumptions, the formulas of Eyring are in good agreement with the phenomenological thermodynamics[13]. In both cases theory and experiment are in good conformity.

2. Statistics of Thermal Vibrations

The most important assumptions for the applicability of thermodynamical statistics is the independence of the particles from one another and the absence of interchange effects between them. Boltzmann — as well as Bose- and Fermi-statistics consider individual particles without interaction. In the gaseous state, photons, electrons as well as molecules coexist. In applying these theories to condensed phases, the individual particle is to be considered, according to Schrödinger[15], either in a continuous medium otherwise the interaction must be taken into account.

We have proposed considering the thermal oszillations of single quasicubic particles and the calculation of the distribution of their vibration energy according to Boltzmann. By superposition of some thermal oszillations we obtain the probability for the displacement of molecules. Moreover, the value for the frequency ν_{th} of the thermal vibration acting in the direction of possible molecular dislocation may be considered in agreement with Eyring having the constant value $\nu_{th} = kT/h \approx 10^{13}$ Hz. Calculations with this average frequency show good agreement with the "äquipartition" theorem and the law of Liouville. One-dimensional vibrating processes are the subject of a publication by Baur[16]. At the lowest position of their respective potential plane independently vibrating particles reach a maximum velocity. We compare this velocity with that of gas molecules and use the well known distribution for the velocity in the gaseous state. The kinetic energy at the minimum is equal to the potential energy reached at the inversion point. Overcoming processes take place, if this potential energy of the oscillating particle is greater than the fixed chemical energy. The same considerations are true for rotation processes. In all cases two or more vibration and rotation states are superposed and act together to overcome a given potential barrier. Here are some examples:

1. First we consider a molecule AB vibrating in the direction $(1) \longleftrightarrow (2)$ (Fig. 1) and having the two possible conformations (a) and (b). The flip-flop process $(a) \longleftrightarrow (b)$ is characterized by a distortion of the angles ϑ, ψ, φ between these two stable stages. Vibrations in the direction $3 \longleftrightarrow 4$ help to reduce these distortions.

2. A free rotating of the CH_3-group is hindered by molecular forces caused by the molecular chain AB (Fig. 2). It is less hindered if cross vibrations of the molecular chain take place simultaneously.

3. The rotation of a segment 1, 2, 3, 4, 5, 6 to the position 1, 2, $3'$, $4'$, 5 is facilitated if the other molecules (Fig. 3) vibrate simultaneously in the direction II $\longleftrightarrow$ III.

In many cases we can describe the situation by a schema (Fig. 4) showing the potential energy between the positions (B) and (B'). The improbable conformation (c) is characterized by a distortion of the valence angles and repulsive forces acting on neighboring molecules. The calculations given in the appendix (I) are limited to one space component in the direction between the two possible conformations. The probability of the transfer of a molecule from the conformation (B) into a second possible position (B') will be favored not only by a large amplitude of oscil-

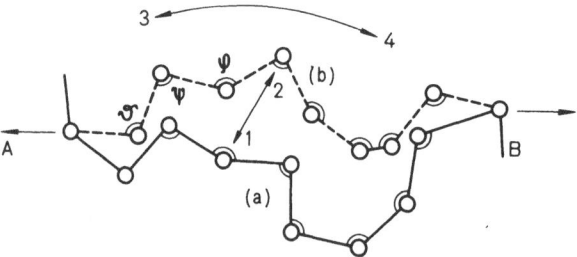

Fig. 1. Molecular dislocating between two possible conformations (a) and (b)

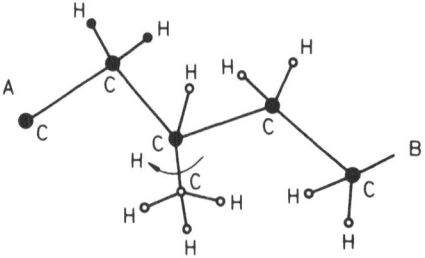

Fig. 2. Hindered rotation of a CH_3-group

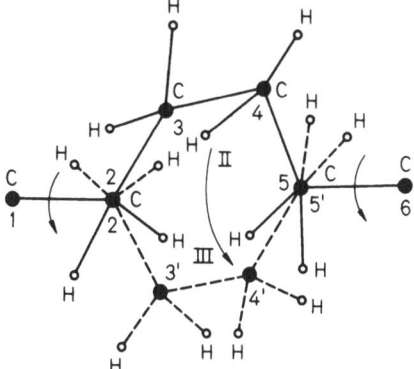

Fig. 3. Segment rotation leading to a chain
molecule from Position II to III

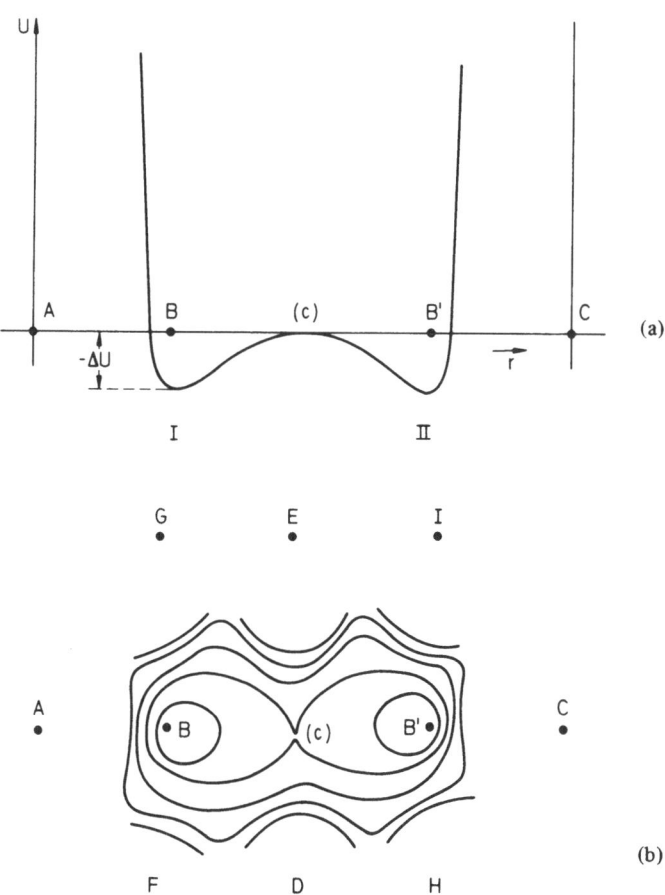

Fig. 4. Cross section of a potential hyperplane in the direction of possible conformations (a)
and topology (ground plane) (b)

lations in the respective direction, but also by random movement of hindering neighbors in the sense that a transition from (B) to (B') is favored. These not-only-but-
also probabilities lead to multiplicative interrelations of the individual probabilities.
The probability of several oszillations collectively possessing a certain energy value
U is proportional in the thermodynamical sense to the number of states characterized by the condition $\sum_i U_i = U$.

The probability for a flip-flop process is then obtained by integration over all cases
in which the total kinetic U of several participating particles exceeds a certain value
ΔU. In the case of superposition of two oscillations influencing one molecular dislocation, we obtain the well known Arrhenius equation

$$W_p = \exp(-\Delta U/kT) \tag{1}$$

It is important to know that this generally applied formula is connected with the
vibration of two particles: one creating the energy hyperplane, the other swinging
in this plane; or in an abstract sense: two thermal oscillations superposed on one
another.

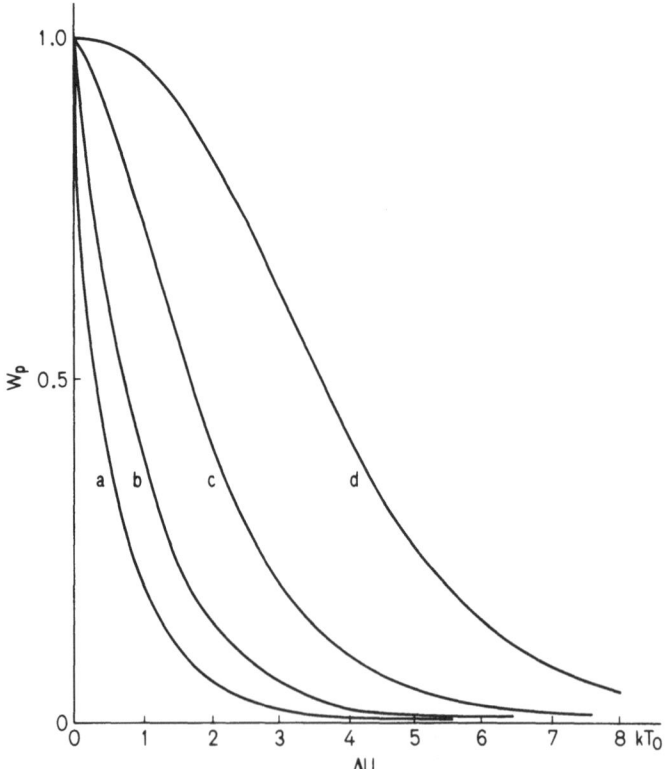

Fig. 5. Probability for molecular dislocations as a function of the activation energy. a) individual
vibrating particles, b) superposing of two thermal vibrations (Arrhenius), c) superposing of four
thermal vibrations, d) superposing of six thermal vibrations

The Arrhenius formula results from the distribution of thermal vibrations in one fixed direction. Zaeschmar[17] obtains more complicated relations.

For the case where vibrations caused by the next neighbors have full influence, we can write the probability for flip-flop processes determined by 4 oscillations as

$$W_p = (1 + \Delta U/kT)\exp(-\Delta U/kT) \tag{2}$$

and for that of 6 thermal vibrations

$$W_p = (1 + (\Delta U/kT) + (\Delta U/kT)^2/2!)\exp(-\Delta U/kT) \tag{3}$$

If the distant neighbors are involved only to a degree represented by a share rate $\beta\Delta U$ with $0 \leqq \beta \leqq 1$, we then get (see the appendix for 4 thermal vibrations)

$$W_p = \frac{1}{1-\beta} \cdot \left(\exp\left(\frac{-\Delta U}{kT} \right) - \beta \exp\left(\frac{-\Delta U}{\beta kT} \right) \right) \tag{4}$$

leading to Eq. (1) for $\beta = 0$ and Eq. (2) for $\beta = 1$.

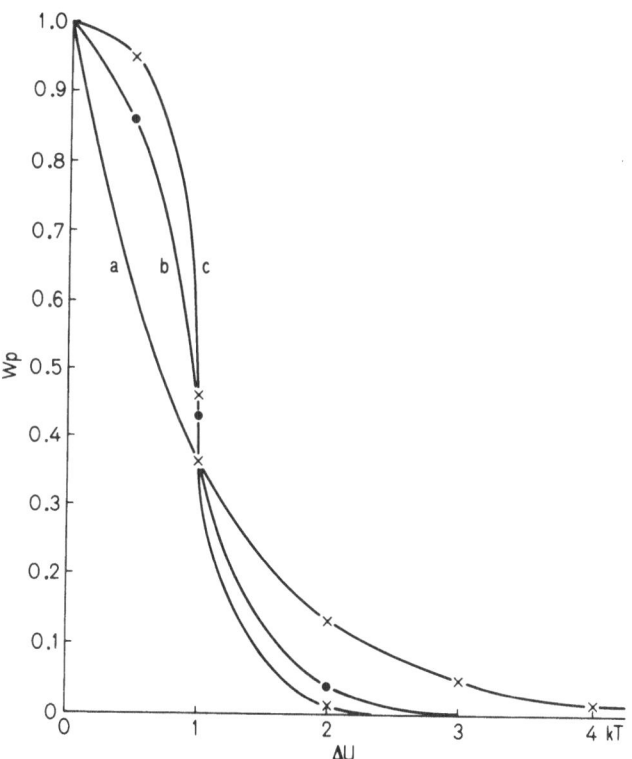

Fig. 6. Probability for molecular dislocations when only a share rate of the thermal vibration energy is localized at the critical point for molecular dislocations. a) Arrhenius formula, b) four independent oscillations, c) six independent oscillations

In Fig. 5 this probability is plotted for a free vibration (Gauss distribution) (a) for 2 coupled vibrations (Arrhenius formula) (b), for 4 vibrations (c) and for 6 vibrations (d). The picture will be completely changed, if we assume that only a share value 1/4 of the total energy of the thermal oscillations influences a special barrier process (b) in Fig. 6. Calculating the same for 6 oscillations superimposed on one another, we obtain curve (c) in Fig. 6.

From Fig. 6 we see: The probability of overcoming a potential barrier (*e.g.* $\Delta U = 2\,kT$) will be reduced if this process is stimulated by more and more thermal vibrations. If we calculate in all cases solely with the Arrhenius equation, we find the activation energy ΔU to be too high (factor 2 to 4).

In spite of this we generally use the Arrhenius equation knowing that under certain circumstances the activation energies thus calculated will be too high. The effects of superposition of thermal vibrations to obtain the probability for barrier processes was also given by Fowler and Guggenheim[18]. These authors verified our results[5] leading to the same formula:

$$W_{\mathrm{p}} = kT/h\left[1 + \sum_n \frac{1}{n!}\left(\frac{\Delta U}{kT}\right)^n\right]\exp - \Delta U/kT \quad \text{[18]}.$$

3. The Activation Energy ΔU and Its Influence on Elastic Behavior and Thermal Expansion

3.1. The Binding Forces in the Lennard-Jones Potential

All molecular dislocations are connected with the overcoming of potential barriers. The height of these depends on the molecular forces. We have to distinguish between intra- and intermolecular forces. The Schrödinger differential equation can be solved for exchange forces in overlapping electronic systems, the so-called atomic orbitals[19]. We obtain almost the same results with empirical formulas used to calculate rotating processes and intramolecular movements, whereas intermolecular fields of force are much more complicated. In simple cases we find spheric symmetrical fields *i.e.* $F = -z\,e^2/(4\,\pi\,\epsilon_0\,r^2)$ for Coulomb forces and $F = -4\,\mu_1^2\mu_2^2/(r^7\,kT)$ for dipoles in random position with respect to one another. Translational motion of macromolecules in general is hindered simultaneously by angular deformation and by the repulsing forces caused by neighboring molecules. In the present situation it is impossible to give exact calculations. In general we use experimental results for example by measuring the heat of vaporization. Moreover, we pay particular attention to the fact that the intermolecular field of force decreases with the distance between the partners in a different manner. The molecular exchange forces have a short range, whereas Coulomb and dipole binding forces are still effective at greater distances.

We use a quasicubic model to try to describe the dependence of the flowing units on the distance, while using a Lennard-Jones potential. We use $n = 6$ for the attractive and $m = 12$ for the repulsive forces. The numeral agreement with the experiments resulting from these assumptions may not be overrated.

For the cohesive energies, the values given by Krevelen[14] are recommended.
We use:

Dispersion potential $\Delta U \equiv 1-4$ kcal/gmol   Polyethylene 2,3 kcal/gmol
                                                         Polypropylene 3,4 kcal/gmol
Dipole binding potential $\Delta U \equiv 4-8$ kcal/gmol Polyvinylchloride 4,2 kcal/gmol
                                                         Polyvinylacetate 6,1 kcal/gmol
                                                         Polymethyl methacrylate 7,2 kcal/gmol
Hydrogen (potential) bonds $\Delta U \equiv 6-20$ kcal/gmol
                                                    Polyethylene terephthalate 14,3 kcal/gmol

We distinguish between intermolecular and intramolecular forces. In most cases
some forces are coupled and must be overcome simultaneously. In the gaseous state
we have to consider single coupling forces between small molecules. The same forces
are also in effect between macromolecules. In this case however, there are a multi-
tude of binding forces which must be overcome simultaneously to separate the molecules. This is one reason why polymers are either solids or liquids. Therefore, in our
quasicubic model described by the Lennard-Jones potential, we can expect average
energy differences of 7–15 kcal/mol for segment dislocations and kink processes.

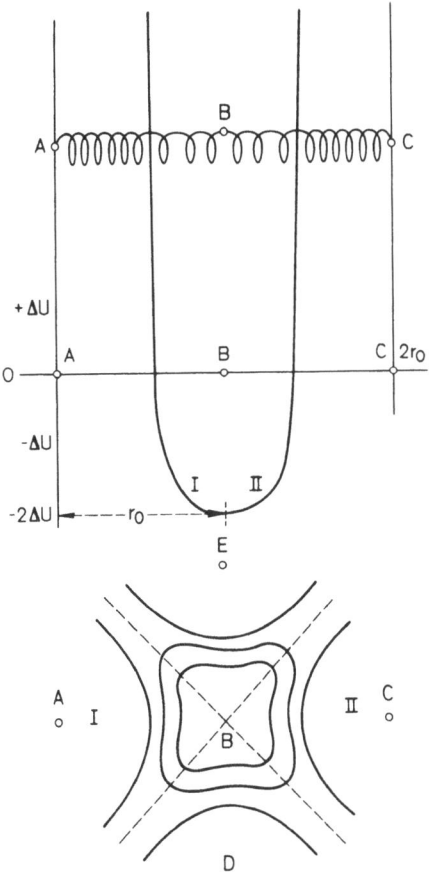

Fig. 7. Potential path for a particle vibrating in a
quasicubic lattice (molecular dislocations are
impossible or improbable only in the direction
of the dashed lines)

These valuations enclose some single molecular adherent places. Thus we calculate an average diameter for these flowing units of about 10^{-7} to 10^{-6} cm.

The necessity to overcome some chemical forces simultaneously increases the activation energy.

We use a given Lennard-Jones-potential for flowing units (that is for segments or kinks) and find in the case of three particles A, B, C.

$$U = \Delta U_0[(r_0/r)^{12} - 2(r_0/r)^6 + (r_0/(2\,r_0 - r))^{12} - 2(r_0/(2\,r_0 - r))^6] \qquad (5)$$

where r_0 determines the minimum position. For $r = r_0$ we get $U = -2\ \Delta U$. The calculations were done assuming a quasicubic structure and an average distance r_0 between the particles (one-dimensional considerations). In this case a particle B is situated between two particles A and C and is characterized by anharmonic thermal vibrations between (I) and (II) (Fig. 7). The repulsive forces give for the potential the terms $(r_0/r)^{12}$ and $[r_0/(2\,r_0 - r)]^{12}$ and the attractive ones are described by the terms $2(r_0/r)^6$ and $2[r_0/(2\,r_0 - r)]^6$. There exists only one possible conformation. In this case flip-flop processes are impossible.

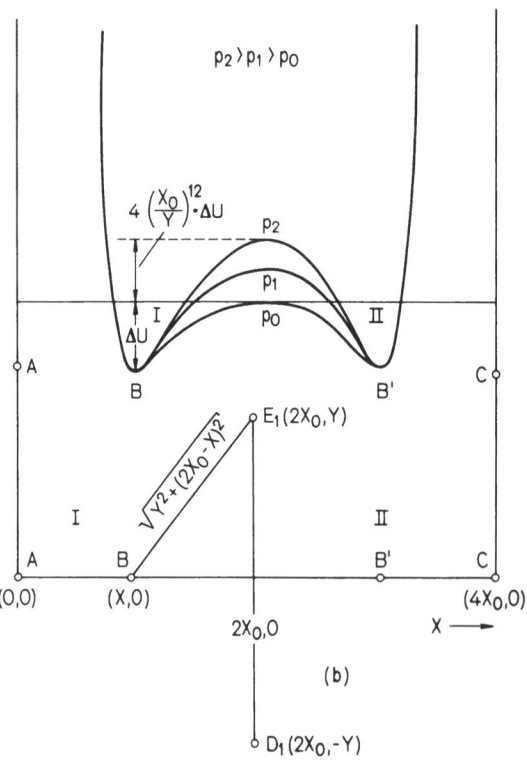

Fig. 8. Energy hyperplanes defined by potential given in Eq. 6 (lateral particles increase the potential barrier *e.g.* with growing pressure (p_1, p_2) the potential may be calculated from the ground plane (b)

Another potential function is given by the following equation (Fig. 8):

$$U = \Delta U[(r_0/r)^{12} - 2(r_0/r)^6 + (r_0/(4\,r_0 - r))^{12} - 2(r_0/(4\,r_0 - r))^6]$$ (6)

In this case the particles A and C are at a distance $4\,r_0$ of one another. There are two potential minima separated by the potential barrier ΔU. For $r = r_0$ we get the energy $U \approx -\Delta U$ and for $r = 2\,r_0$, $U \approx -1/16\ \Delta U \approx 0$.

Since there are two conformations with the same probability particle B can have either the Position I or II.

3.2. The Influence of External Pressure on the Activation Energy ΔU

The quasicubic model does not describe the observed dependence on pressure of all deformation processes. We therefore introduce a spatial model (Fig. 8) in which the considered flowing units lie in the xy plane. We calculate the change of the potential for a particle going from Position I to II. In addition to the particles $A\,(0, 0), B(x, 0)$ and $C(4\,x_0, 0)$ we introduce the lateral particles $D(2\,x_0, -y), E(2\,x_0, +y), F(2\,x_0, z)$ and $G(2\,x_0, -z)$ with F above and G below the xy-plane and $|z| = |y|$. The potential for the particle B with four lateral particles, is

$$
\begin{aligned}
U = \Delta U[&(x_0/x)^{12} - 2(x_0/x)^6 + (x_0/(4\,x_0 - x))^{12} - 2(x_0/(4\,x_0 - x))^6 \\
&+ 4\,x_0^{12}/[y^2 + (2\,x_0 - x)^2]^6 - 8\,x_0^6/[y^2 + (2\,x_0 - x)^2]^3]
\end{aligned}
$$ (7)

where x_0, x, y, and $\sqrt{x_0^2 + y^2}$ describe distances without any vectorial character [$x_0 = r_0$ in Eqs. (5) and (6)].

With $x = x_0$ we arrive at $U \approx \Delta U[4\,x_0^{12}/(y^2 + x_0^2)^6 - 8\,x_0^6/(y^2 + x_0^2)^3 - 1]$ and for $x = 2\,x_0$ the expression $U \approx \Delta U[4\,x_0^{12}/y^{12} - 8\,x_0^6/y^6]$ results.

In the case $|y| < |x_0|$ we calculate the height of the potential barrier which must be overcome to be approximately (Fig. 8)

$$U_{ov} \approx \Delta U(1 + 4(x_0/y)^{12})$$ (8)

The distance y is a function of the pressure. Increasing the distance between the disturbing lateral particles leads to $U_{ov} = \Delta U$. This is true for $y > r_0$.

3.3. Approximate Calculation of the Thermal Expansion in the Lennard-Jones-Model (Temperature Dependence of the Activation Energy)

In some publications[2, 14, 20, 21] the activation energy ΔU is defined as depending on the temperature and is calculated using an increment method. In our case ΔU is given by the chemical binding forces and is influenced only by the thermal expansion.

To prove the influence of T on ΔU we return to our one dimensional model given by a Lennard-Jones potential. We consider particle B to be fastened between A and C by a nonlinear spring (Fig. 7). The force F is given by

$$F = -\frac{\mathrm{d}U}{\mathrm{d}r} \tag{9}$$

As the repulsive and attractive forces are not equal, the repulsive forces will prevail in the direction $A \leftarrow B$ if B is in the neighborhood of A and in the direction $B \rightarrow C$ if B is near C. In this manner an average force results increasing the distance $A - C$, caused by thermal vibrations. It is easy to calculate the excess of the repulsive forces [Eqs. (5) and (6)] in the direction $A - C$ using a Taylor development which includes the third term. We find, using $r = r_0(1 + \Delta r/r_0)$ from Eq. (5)

$$F = 756 \, \Delta r^2 \Delta U/r_0^3 \tag{10}$$

with Δr the distance with respect to the equilibrium position r_0. Besides this we can calculate the average slope of the spring characteristic $\dfrac{\mathrm{d}F}{\mathrm{d}r}(r = r_0)$. With Eqs. (5) and (6) we get

$$\frac{\mathrm{d}F}{\mathrm{d}r} \approx -144 \, \Delta U/r_0^2 \tag{11}$$

and $\dfrac{\mathrm{d}F}{\mathrm{d}r} \approx -72 \, \Delta U/r_0^2$ respectively

Since the thermal vibration energy $1/2 \, kT$ agrees with the potential energy

$$\frac{F^2}{2 \, \dfrac{\mathrm{d}F}{\mathrm{d}r_{(r=r_0)}}} \quad \text{(at the inversion point) we find} \quad \frac{1}{2} kT = \frac{F^2}{2 \, \dfrac{\mathrm{d}F}{\mathrm{d}r_{(r=r_0)}}}$$

To a first approximation we have

$$\Delta F = \Delta r \, \frac{\mathrm{d}F}{\mathrm{d}r_{(r=r_0)}} \quad \text{and} \quad \Delta r^2 = \frac{kT}{\dfrac{\mathrm{d}F}{\mathrm{d}r_{(r=r_0)}}}$$

The excess of the repulsive forces near the inversion point A and C amounts to

$$\Delta F = 756 \cdot kT \, \Delta U \Big/ \left(r_0^3 \, \frac{\mathrm{d}F}{\mathrm{d}r_{(r=r_0)}} \right)$$

Increasing the temperature to $T + \delta T$ yields $F + \delta F$ so that

$$\delta F = 756 \cdot k\Delta U \cdot \delta T \Big/ \left(r_0^3 \, \frac{\mathrm{d}F}{\mathrm{d}r_{(r=r_0)}} \right) \tag{12}$$

This effect is coupled with an elongation δr

$\delta r = r_0 \, \delta F/(144 \, \Delta U)$
[using Eq. (11)]
As $\delta r = r_0 \cdot \beta_{th} \cdot \delta T$, we find that with Eqs. (11) and (12) the thermal expansion coefficient is

$$\beta_{th} = \frac{7}{192} \cdot k/\Delta U \tag{13}$$

The value of β_{th} depends on the chosen potential function. With $\Delta U = 3$ kcal, we obtain $\beta_{th} \approx 2{,}5 \cdot 10^{-5}$. In reality the value is about $\beta_{th} \approx 6 \cdot 10^{-5}$ to $8 \cdot 10^{-5}$. This means that the anharmonicity of the thermal vibrations is even greater[21]. For the potential Eq. (6) (Fig. 4) we need only to calculate the elastic deformation by thermal vibrations in the chosen approximation for one half wave, if B is situated in the neighborhood of A. This leads to

$$\beta_{th} = 0{,}012 \, k/\Delta U \tag{14}$$

3.4. The Glass Transition

With the same model we can understand the existence of a glass transition effect in amorphous materials and calculate its temperature range. (Also see[11].) Consider Fig. 9, a drawing of the structure of cristalline and amorphous polymers. We make two parallel plane cuts in the material at a distance of $n \, r_0$ from each other. We choose $n \, r_0$ in such a way that when heating the material from zero to the glass transition temperature calculated for the distance $n \, r_0$ an enlargement r_0 will take place. With $T_g = 400 \, K$ and $\beta_{th} \approx 8 \cdot 10^{-5}$ we find $n \approx 30$ ($n \, r_0 \approx 0.5 \; 10^{-5}$ cm, the

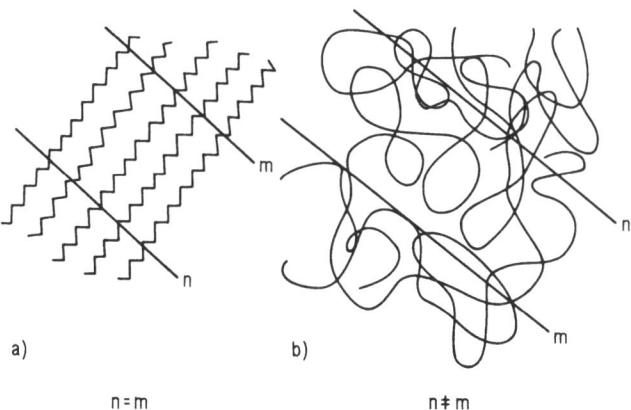

a)   b)

n = m   n ≠ m

Fig. 9. Fluctuation of the density in cristalline (a) and amorphous (b) material (n and m represent cross section of the molecules)

wavelength of light e.g. is $0.5 \; 10^{-4}$ cm). We find n cuts through the molecules of one plane and m for that of the other $(n \neq m)$. Wir respect to cristalline pieces, the number of cuts for both planes is approximately equal.

In the amorphous pattern we observe a spatial fluctuation of density as well as different repulsive and attractive forces. Describing the amorphous structure with $(n - m)/n = \chi$ and using the 6–12-potential of Eq. (5) we obtain an internal shearing stress

$$\sigma = (n - m)/n \cdot \beta_{th} \Delta T E_m$$

or with Eqs. (13) and (11)

$$\sigma = 7/192 \cdot k(n - m)/n \cdot \Delta T \cdot E_m/\Delta U = \frac{7 \cdot 144}{192} \; k\chi \Delta T/r_0^3 \tag{15}$$

($\chi = 0$ for crystalline material and $\chi = 1$ for the highest state of disorder).

The energy necessary to form a hole is approximately $\sigma r_0^3 = \Delta U$. This leads to

$$\Delta T \approx T_g \approx \frac{4}{21} \cdot \frac{\Delta U}{k\chi} \tag{16}$$

With $\Delta U = 2$ kcal/gmol we find $T_g \approx 200 \; K$ for $\chi = 1$ (complete state of disorder) and $T_g \approx 400 \; K$ for $\chi = 0.5$. In the case of partially crystalline material, the glass transition temperature exists only for the amorphous part.

Heating up amorphous solids, we observe an unsteady increase $\Delta \beta_{th}$ of the thermal expansion coefficient β_{th} in a characteristic temperature range. This effect is coupled with the formation of the free volume (Fig. 10) and a strong decrease of

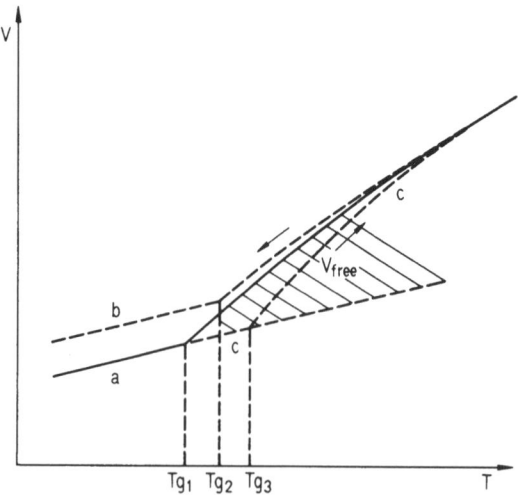

Fig. 10. Volume of amorphous materials near the glass transition temperature. a) very slow heating and cooling: equilibrium state (T_{g1}), b) quick cooling (T_{g2}), c) quick heating (T_{g3}), V_{free} = free volume

the viscosity ($\eta \approx 10^{13}$ Poise at T_g). Moreover, this temperature is characterized by the beginning of molecular mobility (β relaxation). This is the phase transition temperature from the glassy to the liquid state.

The formation of holes, that is transformation from the potential of Eq. (5) to the potential of Eq. (6) in our model, may be supported by thermal vibrations. If we heat up slowly, meaning that we wait long enough, the transition temperature can be found at a lower temperature range (Fig. 10). The glass transition temperature is reduced by annealing and increased by heating (curve c) or cooling (curve b) quickly. Equation (16) gives an upper limit. The curve a in Fig. 10 gives the equilibrium state. The measured transition temperature T_g will shift to a higher temperature range (some degrees) if the heating is performed at a higher speed (curve c). We observe a relaxation phenomenon forming the equilibrium volume at each temperature value.

We summarize: The glass transition process is caused by fluctuations of the density in amorphous material thereby creating shearing stresses by thermal expansion in districts with about 10^{-5} cm diameter and forming holes of about atomic size. In an agreement with Jenckel[23] Breuer and Rehage[24] we consider this perturbation of thermal equilibrium and the relaxation of volume and molecular mobility[22, 25] as a freezing process. It is connected to the creation of the free volume (V_{free} in Fig. 10).

In our simple theory we neglect the influence of different intermolecular forces, leading to different energies for hole formation. This fact gives rise to the increment methods of calculating T_g given by Illers[26] and Becker[27]. Describing glass transition effects as molecular dislocations does not refute the known phenomenological theories of Gordon[28], Kanig[30], Boyer[29] and Kovács[31].

4. The Free Volume and the Rate α of Particles Participating in Molecular Exchange Processes

We define the share rate of the mobile flow units as

$$\alpha = V_{free}/V_{solid} \tag{17}$$

We distinguish: V_0 equilibrium volume at very low temperatures, $V_{solid} = V_0(1 + 3\,\beta_{th}\,\Delta T)$, $V_{total} = V_0(1 + 3(\beta_{th} + \Delta\beta_{th})\Delta T$ and $V_{free} = 3\,V_0\,\Delta\beta_{th}\,\Delta T$.

This means: The share rate α is given in our simple model as the quotient of flowing units, which can be described by the potential of Eq. (6), (fluctuating particles) and the fixed particles [Eq. (5)]. Fig. 11a.

The increase $\Delta\alpha$ of the share rate α begins at the glass transition temperature with the slope $\dfrac{d\alpha}{dT}$ or $d\alpha = 3\,\Delta\beta_{th}\,dT$.

$\Delta\beta_{th}$ is the jump of the linear thermal expansion coefficient at the glass transition temperature. As Fig. 11b shows, $d\alpha$ is proportional to the rate $(1 - \alpha)$ of the flowing units, that is to the molecular arrangements, which can form some confor-

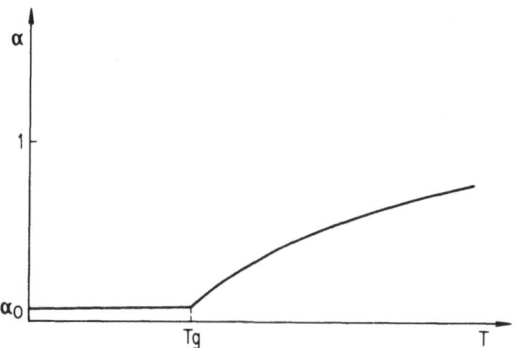

Fig. 11a. Share rate α for mobile segments or flowing units

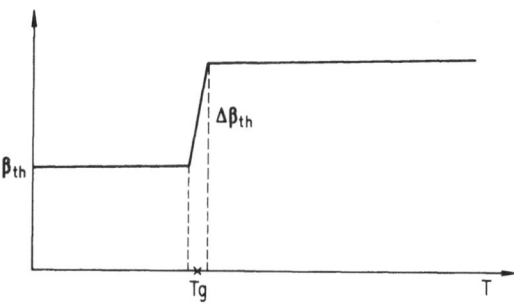

Fig. 11b. Change of the thermal expansion coefficient β_{th} at the glass transition interval (b)

mations. The possibility for this increases with the temperature. We get the differential equation

$$d\alpha = C_1(1 - \alpha)d\Delta T$$

We begin the temperature scale ΔT at the glass transition temperature (T_g) $(C_1$ is a constant). The result of the integration is

$$\ln(\alpha - 1) = -C_1\Delta T + C_2$$

or with $\Delta T = T - T_g$

$$\alpha = [1 - (1 - \alpha_0)\exp - C_1(T - T_g)]$$

In this case the share rate α at the glass point T_g is $\alpha = \alpha_0$ and will be $\alpha = 1$ for $T - T_g \to \infty$.
Since

$$\frac{d\alpha}{d\Delta T} = 3 \cdot \Delta\beta_{th} \text{ for } T = T_g, \text{ we have}$$

$$C_1(1 - \alpha_0) = 3 \cdot \Delta\beta_{th} \text{ and } 3 \cdot \Delta\beta_{th} \approx C_1; \alpha_0 \ll 1.$$

We get

$$\alpha = 1 - (1 - \alpha_0)\exp(-3 \cdot \Delta\beta_{th} \cdot \Delta T) \tag{18}$$

for

$$T > T_g$$

On the other hand; $\alpha = \alpha_0$ for $T \leqq T_g$ (Fig. 11).
This equation is quite similar to the function

$$\alpha = \exp(-C_3/(C_4 + T - T_g) \tag{19}$$

for $T > T_g$
From this we can calculate

$$\alpha = \exp(-C_3/C_4 + C_3/C_4 \cdot (T - T_g)/(C_4 + T - T_g))$$

or

$$\alpha = \alpha_0 \exp(C_3/C_4 \cdot (T - T_g)/(C_4 + T - T_g)) \tag{20}$$

with

$$\alpha_0 = \exp(-C_3/C_4)$$

In this we can substitute

$$a = \exp(-C_3/C_4 \cdot (T - T_g)/(C_4 + T - T_g))$$
(Williams-Landel-Ferry (WLF)[2])

We find $\alpha = \alpha_0/a$. In Fig. 12 the broken line (a) represents the Eq. (20) or the (WLF) whereas curve (b) corresponds to Eq. (18).

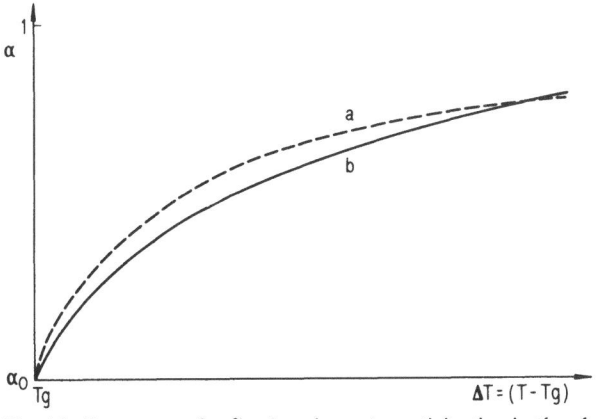

Fig. 12. Share rate α for flowing elements participating in the place exchange processes
WLF function [curvature (a)]
statistical theory given in this work [curvature (b)]

Note: The formation of holes is connected to the absorption of energy (endo-thermic process) and with a jump in the course of the specific heat c_p. Likewise the stimulation of coupled place exchange processes (Chapter 9) (gradually transition to single molecular dislocations with growing free volume) increases the specific heat (Fig. 11) (more degrees of free motion). For each molecular hole we expect 4 to 6 new possible molecular conformations and a lot of molecular displacements giving us an observable macroscopic deformation. After producing many holes by heating, the remaining stress leads to molecular dislocations. The dependence upon tempera-ture will be described, using the WLF function of Eq. (20) and the similar Eq. (18). That means: In the neighborhood of the glass transition temperature Eqs. (18) and (20) describe the temperature behavior.

5. The Elastic and Dielectric Potential

An external shearing or tensile stress creates an elastical deformation even in the internal structure of a body. We observe an elastic potential, influencing the height of the potential barriers given by atomic forces and changing the free motion of the molecules and molecular segments. The exact calculation consists of considering the Hamiltonian operator including external stresses and solving the Schrödinger equa-tion with a perturbation calculation.

Here a classical approach will be sufficient.

An external stress changes the thermodynamic equilibrium related to the distri-bution of the mobile molecular segments among the possible conformations (Fig. 13). In general the sequence for a molecular dislocation will be triggered by the absorp-tion of thermal energy and concluded after the emission of the same. We calculate a change A relative to the next saddle point of the energy hyperplane caused by an external stress σ at the position of the mobile part of a molecule. Using the law of Hooke as well as the average distance r_0 between the flowing units and a force $F = \sigma r_0^2$ we get the energy

$$A_{el} = \int_{\sigma_1}^{\sigma_2} \frac{\sigma r_0^3 d\sigma}{E_m} = \frac{\sigma_2^2 r_0^3 - \sigma_1^2 r_0^3}{2 E_m} \tag{21}$$

We set $\sigma_2 = \sigma_1 + \Delta\sigma$ where $\Delta\sigma$ is the decay of the stress if a particle move from the minimum of U to the saddle point (distance r_0) (Fig. 13). With $\Delta\sigma \ll \sigma_1$ and $\sigma_1 = \sigma$

$$A_{el} \approx \frac{\sigma \cdot \Delta\sigma \cdot r_0^3}{E_m}$$

With $\sigma_2 = 0$ we get $A_{el} = \text{Const } \sigma^2$ (leading to the formula of Ostwald de Waele). The next n flowing units (segments or molecules) will also change their elastic poten-tial due to one single molecular displacement so that the total energy transfer is given by $A_{el} = n A_{single}$. The shift r_0 is related to n flowing units and with the law of Hooke we can write

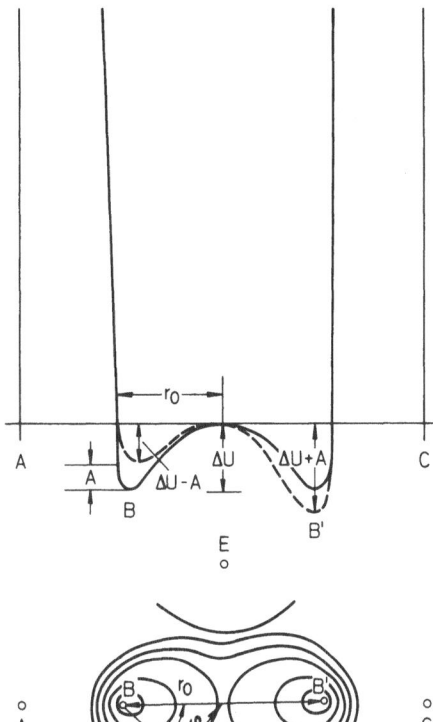

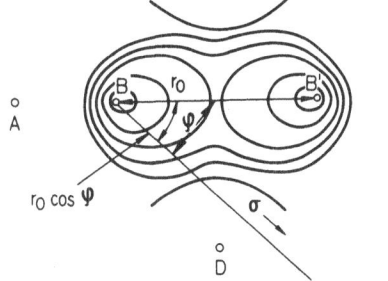

Fig. 13. Potential hyperplane and its variation A created by an external stress σ (φ is the angle between external stress and the direction of possible place exchange processes

$$\frac{\Delta\sigma}{E_\mathrm{m}} = \frac{r_0}{n\, r_0} \qquad (22)$$

$$A_\mathrm{el} = \frac{n\, r_0 \cdot \sigma \cdot r_0^3}{n \cdot r_0} = \sigma r_0^3$$

In our molecular model we use a modules E_m for the molecular dimensions.

This elastic energy will be transformed into thermal energy, that is, we observe the stimulation of thermal vibrations. All flowing processes therefore constitute a transfer of elastic (or dielectric) energy into thermal energy. As we see in Fig. 13 between the direction of an external stress σ and that of the possible motion of a flowing unit we have an angle φ. The distance between the minimum and the saddle point will be r_0. The distance related to the direction of the stress σ is $r_0 \cos \varphi$. Therefore we external force from B to B' will be $F = \sigma r_0^2 \cos \varphi$.

As all angles have the same probability, we have

$$A_\mathrm{el} = \sigma_0 r_0^3 \int_0^{\pi/2} \frac{\cos^2\varphi \sin \varphi \, 2\,\pi \, d\varphi}{2\,\pi} = \sigma r_0^3/3 \qquad (23)$$

This equation may be changed to $A_{el} = \sigma r_0^2 b/3$, if the average distance between the possible conformations is b.

The place exchange effects may also consist of rotary motions of molecular globules or clusters with the radius r_0. In this case a rotation over the angle θ gives the elastic deformation energy

$$A_{el} = r_0^2 \pi \sigma_0 r_0 \theta/3 \tag{24}$$

The 3 in the dominator is caused by the deviation from the direction of the external stress and the possible motion from (B) to (B').

In the following calculation the elastic energy $A = \sigma r_0^3/3$ will be used.

Now let us calculate the alteration of the potential energy U in the case of polarisation by an electric field E_{or}. The orientation field E_{or} depends on the external electric field E which is given by Clausius-Mossotti $E_{or} = (\epsilon + 2)/3 \, E$ or with somewhat better results by Onsager[32]

$$E_{or} = \frac{1}{1 - f_r \alpha} \cdot \frac{3 \, \epsilon_{st}}{2 \, \epsilon_{st} + 1} \cdot E$$

Where E the external electric field

ϵ_{st} the static dielectric constant
α the molecular polarisability

The factor f_r will be given by

$$f_r = \frac{1}{4 \pi a^3 \epsilon_0} \cdot \frac{2 \, \epsilon_{st} - 2}{2 \, \epsilon_{st} + 1}$$

where $2 \, a$ is the diameter of the dielectric unit (sphere) and ϵ_0 the dielectric constant of vacuum.

A more realistic field has been calculated by Fröhlich[33]. Each elementary dipole within a sphere contributes an equal amount to the polarisation created by an external field. Whereas Kirkwood[6] and Fröhlich[33] developed their theories without considering the typical polymer behavior Yamafuji[34] and Ishida[35] noted the possibility of linear chain motion. As neighboring dipoles are coupled, the orientation effects are much more complicated than in the case of single dipoles.

In our model concerning the orientation effects of dipoles we do our calculations using an orientation field given by Onsager[32] whereby the influence of neighboring dipoles is included in the given potential functions U. Although we know that this does not remove the difficulties, the exact calculations of the potential hyperplanes are so complicated that it will be better to do the calculation using some defined barriers $\Delta U_1, \Delta U_2, \Delta U_3$ which lead to α, β, γ relaxation, instead of calculations with special simplified models. Furthermore, there are different distributions. $G(\tau)$ of relaxation times. In general the complex dielectric constant ϵ^* will be calculated according the equation

$$\frac{\epsilon^* - \epsilon_\infty}{\epsilon_{st} - \epsilon_\infty} = \int_0^\infty \frac{G(\tau_i)d\tau_i}{1 + j\omega\tau_i} \tag{25}$$

where ϵ_∞ is the dielectric constant for very high frequency. The relaxation times τ_i depend on the height of the barriers. Fröhlich[33] used an arbitrary rectangular distribution of the relaxation times. In the theory of Fröhlich[33] the activation energy U_i is a strong function of the temperature whereas in our case it only depends on the chemical binding forces. A much more complicated spectrum for the distribution of relaxation times was given by Kirkwood[36]. The broadening of the loss curves due to a distribution of relaxation processes was already observed 1941[5]. As the orientation processes are most evident for molecular dislocations, some authors were primarily concerned with the molecular orientation. They considered thermally activated processes across a potential barrier between two equilibrium potentials. In all cases, the probability of molecular displacements across a given barrier ΔU leads to the Arrhenius formula: $W_p = \exp(-\Delta U/kt)$. As the chance for molecular dislocations per unit of time depends on the number of thermal thrusts, we find the factor $\nu_{th} = h/kT$.

In reality ν_{th} concerns a wide range of thermal vibrations (in the acoustic and optical region).

The first publications on this object were by Kirkwood and Fuoss[7], Holzmüller[5], Fröhlich[33], Yamafuji[34], Ishida[35], Saito[37], Hoffman[38] and Gottlib[39]. Yamafuji and Ishida found good results by mixing the barrier theory and the hydrodynamic theory for viscosity to explain the loss processes. They distinguished the influence of long and flexibly attached dipoles from short dipoles connected to the main chains. In a subsequent publication Yamafuji[40] obtained the rate of flip-flop processes creating the α-dispersion using the Eyring formula $W_p = kT/h \cdot \exp(\Delta ST - \Delta H)/kT$. The model used by Gottlib[39] is also based on two different minima of potential energy separated by a barrier. This model leads to a correlation between dielectric and mechanical losses in the same manner as we arrived in our publications. The theory of Hoffman[38] is concerned primarily with relaxation processes in molecular crystals. It is an increment method calculating special modes ϵ_i for the relaxation. Hoffman using the theory of Onsager for the complex dielectric constant obtains

$$\epsilon^* = \epsilon_{st} + \frac{4\pi\mu_0^2 p}{3kT}\left(\frac{3\epsilon_{st}}{2\epsilon_{st}+1}\right)\left(\frac{\epsilon_{st}+2}{3}\right)\sum\frac{\Delta\epsilon_i}{1+j\omega\tau_i} \tag{26}$$

where p is the degree of polymerization, μ_0 the dipole moment of a side group and N the number of dipole units in one volume unit. The transition rate theory describing species, which relax from one possible position to another was also used by G. Williams[41]. The author makes use of Slater and Guggenheim's theory of unimolecular reactions[42]. A good summary of all theories concerning relaxation processes in polymers is given by McGrum, Read and Williams[43].

We now return to the problem of calculating the change $2A$ of the potential energy U as caused by an external electric field. To find the probability for transition between two possible states (*3*) and (*4*) we calculate the difference for the potential energy (Fig. 14) $\mu_0 E_{or} (\cos \delta_1 - \cos \delta_2)$ and set the z axis in the direction of the electric field thereby forming an angle δ with the dipole. The stable state characterized by a minimum of U and the saddle point $U + \Delta U$ gives the direction of the thermal vibrations leading to a molecular dislocation. Let us consider 3 examples:

1. The dipole moment μ_0 may rotate around an axis (*AB* in Fig. 14). The two stable positions include the angle π and form the angles δ and $\pi - \delta$ with respect to

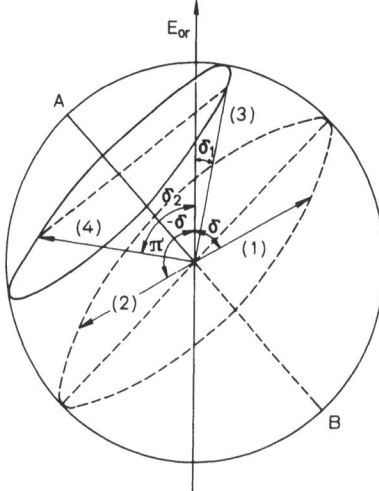

Fig. 14. Orientation polarisation of dipole moments in an electric field E_{or}, (1), (2) possible orientation angles δ and $\pi - \delta$ and (3), (4) stochastic orientation (angles δ_1 and δ_2)

the direction of the electrical field (Fig. 14). For the average change of the potential energy caused by an electrical field, we then calculate:

$$2 A_{diel} = \frac{1}{2\pi} \mu_0 E_{or} \int_0^{\pi/2} [\cos \delta - \cos(\pi - \delta)] 2\pi \sin \delta \, d\delta$$

After integration we have:

$$A_{diel} = 0.5 \, \mu_0 E_{or} \tag{27}$$

2. The dipole may possess two eligible positions marked by the angles δ_1 and δ_2 with respect to the direction of the field. In a previous publication[44] we obtained

$$A_{diel} = 1/3 \, (\mu_0 E_{or}) \tag{28}$$

3. In many cases 3 possible positions exist differing by a solid angle of $120°$ (tetrahedron angle $109°$). We compute the average difference of the dipole energy $2 A_{diel}$ in an electric field to be

$$2 A_{diel} = E_{or} \mu_0 (\cos x_1 - \cos x_2) \quad \text{(Fig. 15)}$$

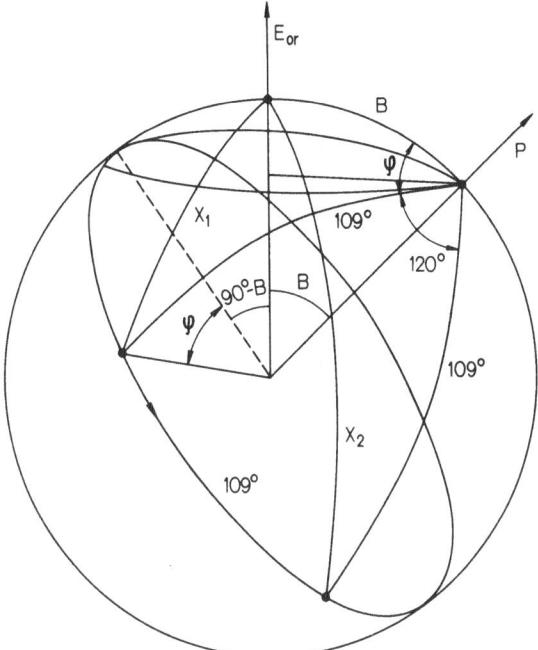

Fig. 15. Dipole orientation in rotating processes (120°, 3 possible positions)

From $\cos x_1 = \cos 109° \cos B + \sin 109° \sin B \cos \varphi$ and

$\cos x_2 = \cos 109° \cos B + \sin 109° \sin B \cos (\varphi - 120°)$

we get $\cos x_1 - \cos x_2 = \sin B \sin 109° [\cos \varphi - \cos(\varphi + 120°)]$. B and φ vary from 0 to $\pi/2$. For a fixed point P (angle B) we must integrate over φ.

$$\overline{\cos x_1 - \cos x_2} = \sin B \sin 109° \int_0^{\frac{\pi}{2}} \frac{\cos \varphi - \cos(\varphi + 120°)d\varphi}{\pi/2}$$

We receive

$$\overline{\cos x_1 - \cos x_2} = \sin B \sin 109° (3 - \sqrt{3})/\pi,$$

where the point P is situated at the solid angle of $2\pi \sin B \, dB$. The integration over B yields the average difference for all angles

$$\overline{\cos x_1 - \cos x_2} = (3 - \sqrt{3}) \text{ sind } 109° \int_0^{\pi/2} \sin^2 B \, dB \cdot 2 \pi/2 \, \pi^2$$

As $\sin 109°$ (tetrahedron angle) is $(2/3) \cdot \sqrt{2}$, then

$$\overline{\cos x_1 - \cos x_2} = 3\sqrt{2} - \sqrt{6}/3 \, \pi = 0.190$$

and

$$A_{diel} = 0.095 \, \mu_0 \, E_{or} \tag{29}$$

In orientated polymers we generally observe a hindered rotation about the direction of the orientation. A_{diel} depends on the direction of the orientation relative to the electrical field[45].

6. The Differential Equations for Reversible and Irreversible Flow Processes and Dielectric Orientation[10, 46]

According to the assumptions about structure with the superposing of thermal vibrations and the elastic energy A caused by external forces we obtain the well known Eyring differential equations[4] for the difference z between the molecular displacements in the direction of the external shearing stress and that of the opposite direction.

Here z_{i0} is the number of flow units per unit area
αz_{i0} the number of mobile segments characterized by two possible positions (1) and (2), z the excess of flip-flop-processes in the direction of an external stress and

$$W_{\Delta U_i} = \exp(-\Delta U_i/kT)$$

the probability of overcoming the barrier ΔU_i (first approximation). If there are two different activation energies ΔU_i and ΔU_k (two different heights of the barriers, Fig. 16), we get:

$$\frac{dz_1}{dt} = \frac{kT}{h} \left[W_{\Delta U_i - A} \left(\frac{\alpha z_{i0}}{2} - z_1 \right) - W_{\Delta U_k + A} \left(\frac{\alpha z_{k0}}{2} + z_1 \right) \right] \tag{30}$$

$$\frac{dz_2}{dt} = \frac{kT}{h} \left[W_{\Delta U - A} \left(\frac{\alpha z_{k0}}{2} - z_2 \right) - W_{\Delta U_i + A} \left(\frac{\alpha z_{i0}}{2} + z_2 \right) \right]$$

Moreover, we have $z = z_1 + z_2$, $W_{\Delta U_i} \cdot z_{i0} = W_{\Delta U_k} z_{k0}$ and $kT/h = \nu_{th} \approx 10^{13}$. As the first equation only depends on z_1 and the second on z_2 the calculation is quite similar to the case (Fig. 13) where

$$\frac{dz}{dt} = \nu_{th} \left[W_{\Delta U - A} \left(\frac{\alpha z_0}{2} - z \right) - W_{\Delta U + A} \left(\frac{\alpha z_0}{2} + z \right) \right] \tag{31}$$

With $W_{\Delta U \pm A} = \exp - (\Delta U/kT \pm A/kT)$ we have the fundamental equation

$$\frac{dz}{dt} = \nu_{th} \exp(-\Delta U/kT) [\alpha z_0 \sinh A/kT - 2z \cosh A/kT] \tag{32}$$

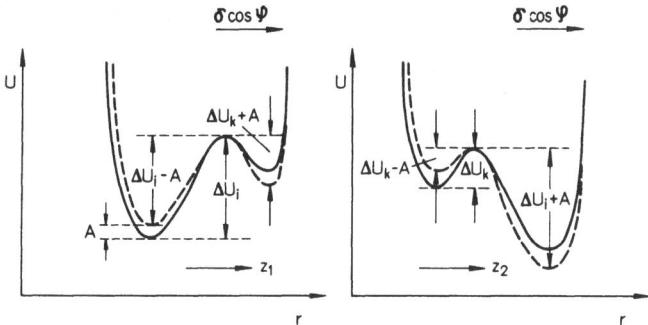

Fig. 16. Potential barriers for molecular dislocations characterized by different activation energies ΔU_i and ΔU_k

Since there are αz_0 mobile elements, in equilibrium we find $\alpha z_0/2$ before and $\alpha z_0/2$ behind the barrier.

The probability was calculated for one thermal vibration. Since there are $kT/h = \nu_{th}$ vibrations per unit of time at first approximation, we get the factor $\nu_{th} = kT/h$ in all cases (h-Planck constant).

Let us refer to the alternative assumption, that the energy hyperplane will remain unchanged by molecular displacements, which means implying that there must be adjacent positions preserving the structure. We speak of reversible flow processes (meaning relaxation processes).

In other cases the formation of a new conformation by molecular dislocation will be characterized by the destruction of the structure in the neighborhood of the moving molecule. This effect makes the possibility of new molecular dislocations more likely. We call these dislocations irreversible displacements. The difference between molecular dislocations in the direction of the external stress and those in the opposite direction does not depend on the number of dislocations z. For irreversible flow processes we get the equation

$$\frac{dz}{dt} = \nu_{th}\alpha z_0 \exp(-\Delta U/kT) \cdot \sinh(A/kT) \tag{33}$$

As is seen in Eqs. (30) to (32) we assume an average number of thermal vibrations $\nu_{th} \approx 10^{13}$.

7. Solutions of the Differential Equations for Flow Processes

There are some characteristic cases with great importance in the field of engineering. As a first approximation we use one dimensional flow processes and consider lateral influences only to describe pressure effects.

7.1. Visco-Elastic Behavior, Relaxation of Deformation Caused by a Constant Stress σ_0

Consequence: Retarded deformation created by gradually occurring displacements favouring the direction of an external stress.

Experimental: Measuring the strain behavior (creep) at a given constant stress. At time $t = 0$ we subject our sample to an external stress σ_0, connected to an elastic potential $A = \sigma_0 r_0^3/3$ at every flow element. This elastic deformation is a disturbance of the thermodynamical equilibrium. At the same time this stress creates a purely elastic deformation $\gamma_0 = \sigma_0/G_0$ of the whole body.

With $A \ll kT$ we can set $\cosh A/kT = 1$ and $\sinh A/kT = \dfrac{\sigma_0 r_0^3}{3\,kT}$, $(\sinh A/kT \approx A/kT)$.

Assuming $T > T_\mathrm{g}$ for reversible flow processes (visco-elastic behaviour) Eq. (32) yields

$$z = z_\infty[1 - \exp(-t/\tau)] \tag{34}$$

with

$$z_\infty = \frac{\alpha z_0}{2}\,\frac{A}{kT} = \frac{\alpha_0 z_0 \sigma_0 r_0^3}{6\,kT} \cdot \exp\frac{C_1(T - T_\mathrm{g})}{C_0(C_0 + T - T_\mathrm{g})}$$

and

$$\tau = \tau_\mathrm{strain} = \frac{1}{2\,\nu_\mathrm{th}}\,\exp\frac{\Delta U}{kT} \tag{35}$$

The relaxation time is

$\tau_\mathrm{strain} \approx 10^{-10}$ to 10^{-9} s (to overcome dispersion forces at room temperature)

$\tau_\mathrm{strain} \approx 10^{-9}$ to 10^{-7} s (for hydrogen bridges at room temperature) and

$\tau_\mathrm{strain} \approx 10^{-6}$ to 10^{+4} s (for multichange processes)

For instance $A \approx 0.3 \cdot 10^{-23}$ Joule yields $\sigma_0 = 1$ N/cm^2 and $r_0 \approx 10^{-7}$ cm, $(kT \approx 4 \cdot 10^{-21}$ Joule at room temperature). The z dislocations per unit area cause a share rate $\gamma = z r_0/z_0 r_0$. Therefore for reversible flow we get

$$\gamma = \gamma_\infty\left(1 - \exp - \frac{t}{\tau}\right) \quad \text{with} \quad \gamma_\infty = \frac{\alpha A}{2\,kT} \tag{36}$$

With $A \approx kT$ we have

$$z = z_\infty\left(1 - \exp - \frac{t}{\tau}\right) \quad \text{or} \quad \gamma = \gamma_\infty\left(1 - \exp - \frac{t}{\tau}\right)$$

with

$$z_\infty = \frac{\alpha z_0}{2}\tanh\frac{A}{kT} \quad \text{or} \quad \gamma_\infty = \frac{\alpha}{2}\tanh\frac{A}{kT} \tag{37}$$

and

$$\tau = \tau_{\text{strain}} = \frac{\exp \Delta U/kT}{2 \, \nu_{\text{th}} \cosh A/kT}$$

(e.g. with $\sigma_0 = 100$ N/cm^2 and $r_0 = 10^{-7}$ cm we get $A \approx 0.3 \cdot 10^{-21}$ Joule). Multi-change processes or molecular dislocations that overcome some barriers simultaneously have to be mentioned here. Finally with $A \gg kT$, setting $\sinh A/kT \approx \cosh A/kT \approx \frac{1}{2} \exp A/kT$ and $\tanh A/kT = 1$ we get

$$\gamma = \frac{\alpha}{2}\left(1 - \exp -\frac{t}{\tau}\right), \quad \tau = \tau_{\text{strain}} \tag{38}$$

with

$$\tau_{\text{strain}} = \frac{1}{\nu_{\text{th}}} \exp \frac{\Delta U - A}{kT} \quad \text{and} \quad \gamma_\infty = \frac{\alpha}{2} \tag{39}$$

All mobile molecules may change their position. The relaxation times decrease exponentially with increasing A. Equation (39) is important in understanding the experiments made to study fatigue, failure ultimate strength and rupture. We have to distinguish between thermal and athermal breaking (ductile and brittle rupture). In the latter case the external stress and the internal elastic potential A are so large ($A \gg U$), that potential barriers can be overcome regardless of the influence of thermal motion. Fracture begins a cracks i.e. where the value for ΔU is low, (weak places in the structure). In general we observe athermal fracture in the low temperature range below the brittle point. The growth of cracks produces new and endangered molecular conformations.

As the crack grows a new surface is being formed. In contrast to this, briefly mentioned athermal rupture, we consider thermal breaking processes according to Eqs. (38) and (39). Since $A = \sigma r_0^3/3$, the relaxation time τ_{strain} depends in an exponential manner on the external stress σ. τ becomes equivalent to the time required for creep rupturing[47].

Flow processes produced by thermal vibrations create new weak spots in the neighborhood of dislocated molecules.

Figure 17 illustrates the dependence of $\ln \tau_{\text{strain}}$ on the external stress and on temperature using Eq. (39). Thermal rupture in all cases is connected to creep processes in the neighborhood. The energy necessary for this flow is much higher than the energy required to form a new surface. Moreover the molecules will be orientated before breaking. This fact is very important in order to understand the increasing strength after orientation. In this case all parallel orientated molecules will be burdened simultaneously[10, 48]. At a high rate of deformation the time is insufficient to create molecular orientation, causing the molecules to break one after another. This is one of the reasons for the observed decrease of the rupture strength in quick deformation processes (brittleness).

In all cases considered hitherto, we calculated, only one relaxation process coupled with one single activation energy ΔU. In reality, we have to consider different relaxation processes happening simultaneously.

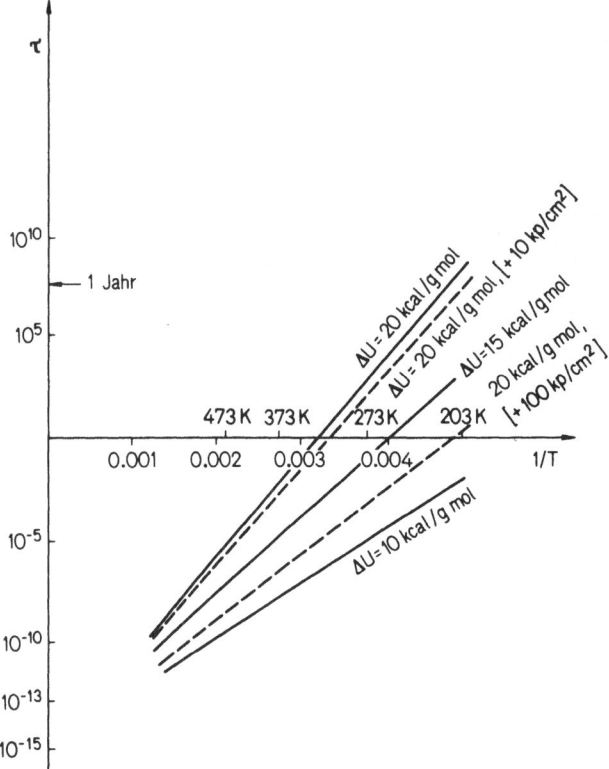

Fig. 17. Relaxation times as a function of the reciprocal absolute temperature for different activation energies (10, 15, 20 kcal/gmol) and different external stresses 10 and 100 kp/cm^2

We have

$\Delta U \approx 1 \cdots 3$ kcal/gmol (low temperature relaxation) or γ process

$\Delta U \approx 3$ to 15 kcal/gmol (kink movement, segment mobility[49]) or β process) and

$\Delta U \gtrsim 15$ kcal/gmol (coupled dislocation processes, mobility of molecular segments fixed near the surface of crystals or α process).

With superposition of these different molecular dislocations there arises a spectrum of relaxation processes leading to the share rate γ.

$$\gamma = \sum_{i=1}^{3} \frac{\alpha_i}{2} \frac{A}{kT} \left(1 - \exp \frac{t}{\tau_i} \right)$$

or

$$\gamma = \sum_{i=1}^{3} \frac{\alpha_i}{2} \tanh \frac{A}{kT} \left(1 - \exp \frac{t}{\tau_i} \right) \tag{40}$$

with

$$\tau_i = \frac{\exp \dfrac{\Delta U_i}{kT}}{2\, \nu_{th} \cosh \dfrac{A}{kT}}$$

where α_i is the share rate of flowing units connected with the α, β and γ process. Moreover ΔU_i does not possess a single value but is broadened to a Gaussian distribution[14].

In contrast to the broad spectrum of activation energies ΔU_i, for calculations we use only a single elastic potential A. The external shearing force causes an unequal internal stress leading to higher values for σ_0 at undangered spots, thus r_0 in Eq. (23) may be lower ($r_0 < 10^{-6}$ cm). We should not forget that every shear or tensile deformation γ must be accomplished by a purely elastic deformation

$$\gamma_{el} = \sigma/G_0$$

The shear modules G_0 or the Young modules E_0 does not depend on time and temperature. Contrary to this defination these modules are often used to describe the whole viscoelastic behaviour including the relaxation processes (Only in these cases the dependence on temperature is sensible).

7.2. Irreversible Molecular Dislocations, Viscous Flow[49]

We have to take the viscous flow processes happening simultaneously into account. For the irreversible molecular movements Eq. (33) yields

$$z_{irr} = \alpha \cdot z_0 \cdot \nu_{th} \sinh A/kT \cdot \exp - \Delta U/kT \cdot t$$

or with

$$\gamma_{irr} = z_{irr}/z_0$$
$$\gamma_{irr} = \alpha \nu_{th} \sinh A/kT \cdot \exp - \Delta U/kT \cdot t \tag{41}$$

Using

$$\alpha = \alpha_0 \exp \frac{C_3/C_4(T - T_g)}{C_4 + T - T_g}$$

and $A = \sigma r_0^3/3$ we find in the case of $A \ll kT$,

$$\gamma = \alpha_0 \nu_{th} \exp \frac{C_3/C_4(T - T_g)}{C_4 + T - T_g} \exp - \frac{\Delta U}{kT} \cdot \frac{\sigma r_0^3 t}{3\,kT} \tag{42}$$

Defining the viscosity η by $\eta = \sigma/\dfrac{d\gamma}{dt}$ with $T > T_g$ we obtain with $\nu_{th} = kT/h$

$$\eta = \frac{3\,h}{\alpha_0 r_0^3} \exp \left(\frac{\Delta U}{kT} - \frac{C_3/C_4(T - T_g)}{C_4 + T - T_g} \right) \tag{43}$$

(with Eq. 20) or

$$\eta = \frac{3\,h\,\exp\,\Delta U/kT}{r_0^3[1 - (1 - \alpha_0)\exp - 3\,\Delta\beta_t(T - T_g)]} \approx \frac{3\,h\,\exp\,\Delta U/kT}{r_0^3(\alpha_0 + 3\,\Delta\beta_{th}(T - T_g))} \tag{44}$$

$$T > T_g$$

(using Eq. 18)
with $\Delta\beta$ = the jump of the linear thermal expansion coefficient at T_g.
   Below the glass temperature T_g we get

$$\eta = \frac{3\,h\,\exp\,\Delta U/kT}{\alpha_0 r_0^3} \; ; (\alpha_0 \ll 1) \tag{45}$$

The slope of the ln η vs. $1/T$ plot is not as steep at the low temperature range.
   For high temperatures (liquid state) we obtain with $\alpha = 1$

$$\eta = \frac{3\,h\,\exp\,\Delta U/kT}{r_0^3} \tag{46}$$

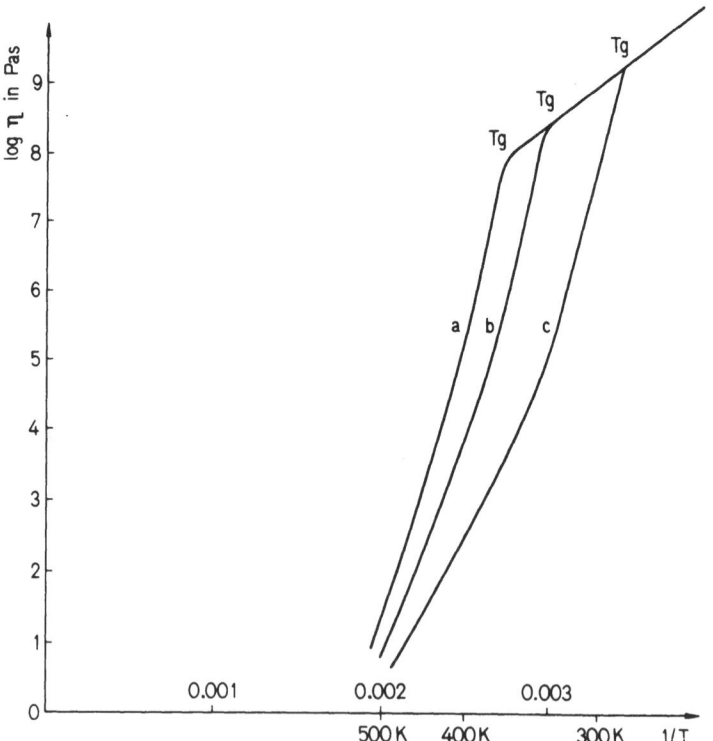

Fig. 18. Viscosity (log η in Pa · s [1 Pa · s = 10 pois]), as a function of temperature for poly-methacrylate (a) PVC (b), and polyvinylacetate (c) according to Eq. (43) (WLF, theoretical curves). The diagram shows the great influence of the free volume near the glass transition temperature

Figure 18 shows the logarithm of the viscosity, calculated on the basis of Eqs. (43) and (45), as a function of the reciprocal absolute temperature. Figure 19 shows the same kind of plot, this time making use of the Eqs. (44) and (45).

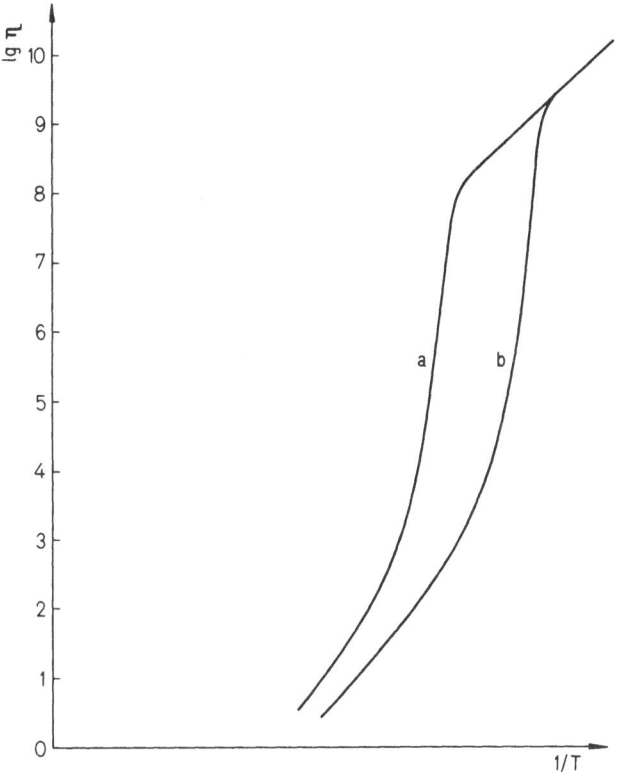

Fig. 19. Viscosity ($\log \eta$ in Pa · s) as a function of temperature for polymethacrylate (a) and polyvinyl acetate according to Eq. (44) (statistical theory, this work)

Since these equations contain the constants C_1, C_2, C_3, C_4 r_0 ($\sim 10^{-7}$ cm) and the constant frequency $\nu_{th} \approx 10^{13}$ of the thermal vibrations, it is easy to find very good agreement with experimental results.

Equation (44) is less arbitrary. We use Refs.[10] and [50] with

$$r_0 = 10^{-9} \text{ m} \qquad \Delta U = 10 \text{ kcal/g} \cdot \text{Mol}$$
$$\alpha_0 = 10^{-7,6} \qquad \nu_{th} = 10^{13}$$
$$\Delta \beta = 15 \cdot 10^{-5} \qquad h = 6,6 \cdot 10^{-34} \text{ Joule sec}$$

and obtain curve a in Figs. 18 and 19.

We can now calculate η/η_{T_g}. With Eqs. (43) and (45) we get

$$\ln \eta/\eta_{T_g} = \left(\frac{1}{T} - \frac{1}{T_G} - \frac{C_3^*/C_4(T - T_g)}{C_4 + T - T_g} \right) \frac{\Delta U}{k} \qquad (47)$$

It should be noted that for $A \ll kT$, η does not depend on the external stress. This quality is typical of Newtonian flow.

With $A \approx kT$ instead of Eq. (44) we get the modified Eyring equation

$$\eta = \frac{\sigma \exp \Delta U/kT}{v_{\text{th}}[1 - (1 - \alpha_0)\exp - 3\,\Delta\beta(T - T_{\text{g}})] \sinh \sigma r_0^3/3\,kT} \tag{48}$$

As $\eta = \sigma/\dfrac{dv}{dr}$ we can integrate this equation for a capillary and for the Couette flow between rotating cylinders (see appendix). In agreement with our previous publication[50] we find the hyperbolic cosine function for the velocity of the current through a capillary. In the case of a Newtonian liquid the velocity is given by a parabolic law. As is seen in Fig. 20 the gradient of the velocity for a non Newtonian liquid is larger near the external regions. Moreover the volume flow rate G through the capillary is about $6\left(\dfrac{A}{kT}\right)^2$ % higher than in the Newtonian case. (The calculation are given in

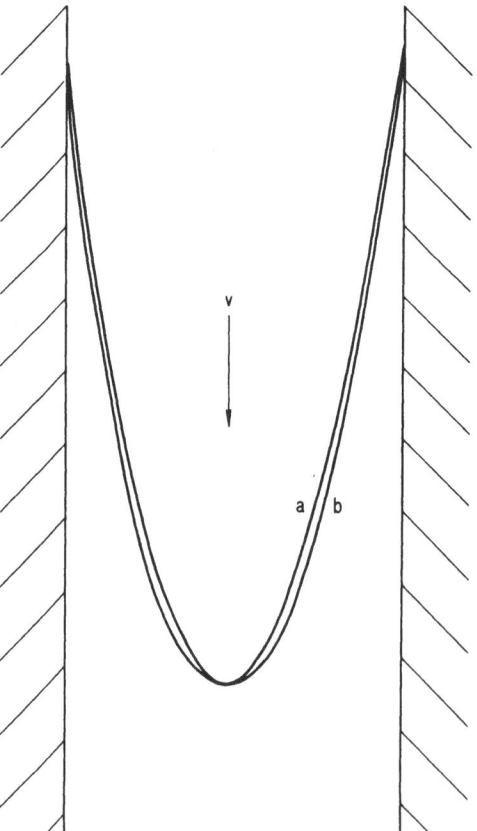

Fig. 20. Velocity v for viscous flow through a capillary. (a) Newtonian liquid (parabolic distribution of the velocity),(b) non-Newtonian liquid (Eyring flow, distribution according to the cosh function)

the appendix and in Ref.[50].) For the behavior of a liquid between two rotating cylinders, the torque is determined to be

$$M = \frac{\eta \cdot 4\pi \cdot L \cdot R_i^2 \cdot \omega}{1 - (R_i/R_a)^2} \qquad \text{(Newtonian liquid)} \qquad (49)$$

$$M = \frac{3\,kT}{2\,\omega r_0^3}\,Ar \sin \frac{2\,\omega \exp \Delta U/kT}{[1 - (R_i/R_a)^2]\alpha \nu_{th}} \qquad \text{(Eyring liquid)}$$

The Eyring flow shows a typical dependence upon stress. With increasing stress the viscosity η decreases (structural viscosity). The calculations are in good agreement with the experimental values. Figures 21 and 22 show the influence of shearing stress on the viscosity.

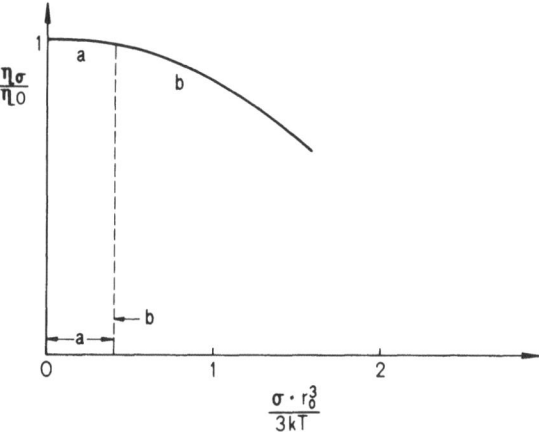

Fig. 21. η depending on stress σ. (a) Newtonian flow, (b) Eyring flow

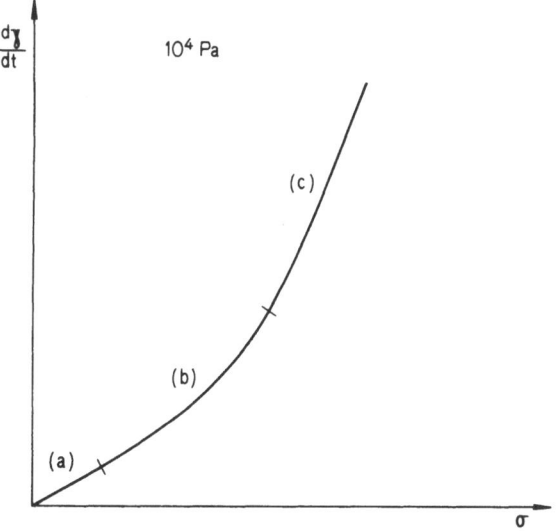

Fig. 22. Shear velocity increasing linearly with stress in the Newtonian range (a) nonlinearly in the Eyring range (b) and rapidly for exponential flow (c)

7.3. Diffusion Processes

We observe also flip-flop-processes of mobile molecular segments that restore the thermodynamical equilibrium in cases, where the distribution of the flowing units does not agree with the statistical distribution of the possible conformations. During the initial stage the number of dislocated particles may be z_1 i.e. after discharge we observe z_1 dislocated flowing units $(t = t_1)$.

We must solve

$$\frac{dz}{dt} = \left(\frac{\alpha z_0}{2} + z_1 - z \right) W_{\Delta U} - \left(\frac{\alpha z_0}{2} - z_1 + z \right) W_{\Delta U} \tag{50}$$

Using $W_{\Delta U} = \nu_{th} \exp - \Delta U/kT$, the number of molecular dislocations in the direction of the thermodynamical equilibrium is found to be

$$z = z_1 \left(1 - \exp - \frac{t}{\tau} \right)$$

with

$$\tau = \frac{1}{2\,\nu_{th}} \exp \frac{\Delta U}{kT} \quad \text{and} \quad z_\infty = z_1 \tag{51}$$

or

$$\gamma = \gamma_1 \left(1 - \exp - \frac{t}{\tau} \right)$$

The return of shear deformation after elimination of an external stress at a moment t_1 is also plotted in Fig. 23 (curve d). This diagram shows the flow processes before and after discharge using Eq. (36) as well as a pure elastic deformation $\gamma_{el} = \sigma/G_0$ and a irreversible flow according to Eq. (41). The elastical deformation γ_{el} disappears immediately after discharge.

Exactly the same mechanism determines diffusion processes. Using the first law of Fick we have a diffusion current.

$$I = -D \frac{dc}{dr} \tag{52}$$

where c is the concentration and D the diffusion coefficient. The place exchange theory starting with Eq. (50) and $W_{\Delta U} = \nu_{th} \exp - \dfrac{\Delta U}{kT}$ leads to

$$\frac{dz}{dt} = 2(z_1 - z)\nu_{th} \exp - \frac{\Delta U}{kT} . \text{ The solution will be} \tag{53}$$

$$z = z_1 (1 - e^{-2t\,\nu_{th}\,\exp\,-\,\Delta U/kT})$$

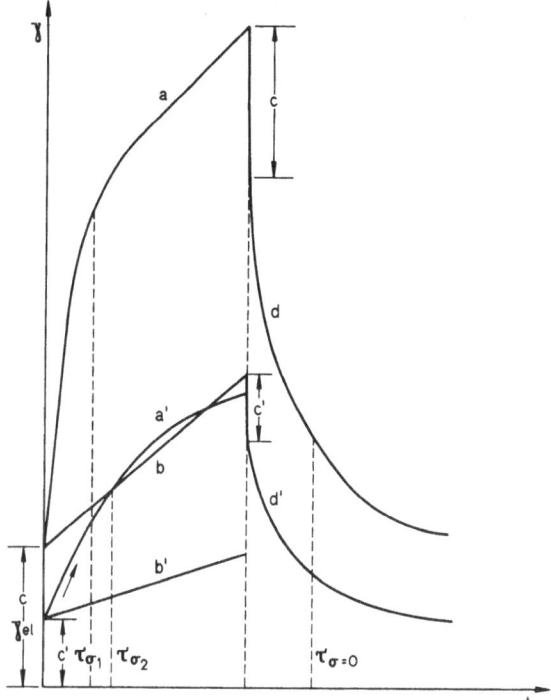

Fig. 23. Deformation and recurrent deformation at constant stress as a function of time, (a) total deformation at high stress (nonlinear behavior, relaxation time τ_{σ_1}), (a') deformation at low stress, (b) viscous flow, (b') viscous flow at low stress, (c) purely elastic deformation for high stress, and for low stress (c'), (d) and (d') recurrent effects (diffusion process)

The diffusion rate $\dfrac{dz}{dt}$ is analogous to $-I$ and the difference $2(z_1 - z)$ is proportional to the gradient of the concentration of dislocated molecules (distance between them in our model, $2 r_0$). We should not forget that z is defined for the number of flowing units per unit area and c is related to the unit volume. With Eqs. (51) and (53) our quasicubic model yields

$$\frac{dc}{dr} = \frac{1}{2 r_0} \cdot \frac{dz}{dr} = \frac{z_1 - z}{r_0^2} \tag{54}$$

and

$$D \cdot \frac{z_1 - z}{r_0^2} = 2(z_1 - z) \nu_{th} \exp - \frac{\Delta U}{kT}$$

so that

$$D = 2 r_0^2 \nu_{th} \exp - \frac{\Delta U}{kT} \tag{55}$$

The diffusion constant D is proportional to the reciprocal relaxation time. We have to distinguish between the following two cases:

1. All molecular dislocations processes are related to the deviation $z_1 - z$.

The deviation $z_1 - z$ decreases according to Eq. 53. As $(z_1 - z) \to 0$ the diffusion rate decreases asymptotically to zero.

2. The deviation from the equilibrium z_1 remains constant. There must be a constant diffusion current stabilizing the concentration gradient. Instead of Eq. (53)

$$\frac{dz}{dt} = 2 z_1 \, \nu_{th} \exp - \frac{\Delta U}{kT}$$

is valid. We then get

$$z = 2 z_1 \cdot \nu_{th} \exp - \frac{\Delta U}{kT} \cdot t \tag{56}$$

This constant diffusion flow according to the first law of Fick can be considered to be analogous to the irreversible flow given by Eq. (41). In Eq. (52) z is a function of time alone. In the case that z depends on space in the sense that the velocity of the molecular dislocations increases with $\frac{d^2 c}{dr^2} \cdot r_0$ we obtain the second law of Fick

$$\frac{\partial c}{\partial t} = -D \frac{\partial^2 c}{\partial r^2}$$

or

$$\frac{\partial c}{\partial t} = -2 r_0^2 \nu_{th} \exp - \frac{\Delta U}{kT} \cdot \frac{\partial^2 c}{\partial r^2} \tag{57}$$

The laws of Fick are concerned with the concentration and the change of concentration with respect to space and time. In this paper we are interested in the number of molecular displacements. If we calculate the change of concentration we must use the laws of Fick. The diffusion constant in all cases is given by

$$D = 2 r_0^2 \nu_{th} \exp - \frac{\Delta U}{kT}$$

Diffusion as a molecular dislocation process has also been mentioned by Müller and Hellmuth[51] and consistently leads to the hopping theory of describing ionic conductivity[52] as a function of the temperature.

The constant diffusion flow according to the first law of Fick, can be considered to be analogous to the irreversible flow given by Eq. (41).

7.4. Fluctuation Rate and Molecular Displacements in the Direction of External Stress

It is interesting to compare the fluctuation rate ν_{fl} for mobile conformations characterized by two or more positions with the molecular dislocations taking place in the direction of an external stress. ν_{fl} is given by $\nu_{th} \exp - \dfrac{\Delta U}{kT}$ and denotes the number of place interchanges in any direction ($\nu_{fl} \approx 10^5 - 10^8 \; s^{-1}$). The number of molecular dislocations in the direction of the external stress, which contribute to the observed deformation, is given by

$$\frac{dz}{dt} = \nu_{th} \exp - \frac{\Delta U}{kT} \left(\alpha_0 z_0 \sinh \frac{A}{kT} - 2z \cosh \frac{A}{kT} \right) \tag{32}$$

Every particle experiences the remanent dislocations per unit time

$$\nu_{rem} = \nu_{th} \exp - \frac{\Delta U}{kT} \left(\sinh \frac{A}{kT} - \frac{2z}{\alpha z_0} \cosh \frac{A}{kT} \right)$$

in the direction of the deformation. The fluctuation time τ_{fl} (time needed to overcome any energy barrier ΔU for any flowing unit) is:

$$\tau_{fl} = \frac{1}{\nu_{th}} \exp \frac{\Delta U}{kT}$$

τ_{fl} is an important quantity used to describe the relaxation processes in broad line paramagnetic spectroscopy (BNMR). The fluctuation time is commonly referred to as correlation time in the NMR literature.

In contrast to this the expectation time τ_{ex} for any molecular displacement contributing to the observed deformation γ is:

$$\tau_{ex} \gtreqless \frac{\exp \dfrac{\Delta U}{kT}}{\nu_{th} \sinh \dfrac{A}{kT}} \tag{58}$$

Its value depends on external stress. Since the relaxation process proceeds, the rate of dislocations in the direction of flow decreases the expectation time τ_{ex} increases.

As $A \ll kT$, τ_{ex} is much longer than the fluctuation time. For $A \approx 0.48 \, kT$ the fluctuation time τ_{fl} has the same value as τ_{ex}.

7.5. Flow Processes Depending on the Pressure p

In the well known equation

$$\frac{dV}{dp} \cdot \frac{1}{V} = - \frac{1}{K} \tag{59}$$

containing the bulk modules K, we can substitute the volume V by using the distance and obtain

$$\frac{da}{dp} \cdot \frac{3}{a} = -\frac{1}{K}$$

Using the approximation Eq. (8)

$$U_{ov} \approx \Delta U \left[1 + 4 \left(\frac{x_0}{y} \right)^{12} \right]$$

after differentiation we get

$$dU_{ov} \approx -48 \, \Delta U \cdot \frac{x_0^{12}}{y^{13}} \, dy$$

[x being the direction between conformation A and C, y the direction between D and E (Fig. 8).]

The introduction of $a = y \, (= z)$ leads to

$$dU_{ov} \approx 4(U_{ov} - \Delta U) \, \frac{dp}{K} \, ,$$

$$U_{ov} \approx \Delta U \left(1 + C_1 \exp \frac{4 \, p}{K} \right)$$

and

$$U_{ov} \approx \Delta U \left(1 + C_2 \, \frac{\Delta p}{K} \right) \tag{60}$$

If we intend to investigate the influence of pressure on the deformation and the relaxation effects, we must replace ΔU with U_{ov}. This is a very rough approximation connected with the 6–12 potential (C_1, C_2 and C_3 are empirical constants, Δp the pressure increase). The relaxation time will then be

$$\tau_{strain} = \frac{1}{2 \, v_{th} \cosh \dfrac{A}{kT}} \exp \Delta U (1 + C_3 \Delta p) \tag{37a}$$

All relaxation and diffusion processes depend on pressure. The viscosity η also increases in an exponential manner

$$\eta = \frac{\sigma \exp[\Delta U (1 + C_3 \Delta p)/kT]}{v_{th}[1 - (1 - \alpha_0) \exp - 3 \, \Delta \beta (T - T_g) \sinh A/kT]} \tag{48a}$$

It is important to note that with increasing pressure the free volume and α decrease. In this way the relaxation time and the viscosity are influenced by pressure to a far greater degree than is calculated using Eqs. (37a) and (48a)[53].

8. Solutions of the Differential Equations for Flow Processes with Variable External Stress and Field

8.1. Relaxation of Stress[11, 12, 54]

We subject a sample of solid polymer material to a sudden deformation process γ_0, with an elastical stress $\sigma_0 = G\gamma_0$ (G shear modules). The original strain gives rise to an increase of the molecular valency angles and the intermolecular distances. From these molecular deformations an elastic molecular potential A_0 arises which in turn causes molecular displacements. These prevail in the direction of the original strain, decreasing the elastic potential A in that neighborhood. We can calculate

$$A = A_0 - Bz, \quad A_0 > Bz \tag{61}$$

(A_0 and B being constants)

In the case that $A \ll kT$ instead of Eq. (32) we get the fundamental differential equation

$$\frac{dz}{dt} = \frac{1}{h}\left[\alpha z_0(r_0^3\sigma_0/3 - Bz) - 2\,zkT\right]\exp - \frac{\Delta U}{kT} \tag{62}$$

The solution of which is

$$z = z_\infty\left(1 - \exp - \frac{t}{\tau_{stress}}\right)$$

with

$$z_\infty = \frac{\alpha z_0 r_0^3 \sigma_0}{6\,kT + 3\,B\alpha z_0}, \quad z_\infty \leqslant \alpha z_0 \tag{63}$$

and

$$\tau_{stress} = \frac{h\,\exp\,\Delta U/kT}{\alpha z_0\,B + 2\,kT}$$

The relaxation time τ_{stress} depends on B and differs from τ_{strain}. In first approximation we can calculate a reduction of the stress due to reversible molecular displacements. According to Hooke we have $\Delta\sigma = G_0\Delta\gamma$ where $\Delta\gamma$ is given by z/z_0. Thus we find

$$A = \sigma r_0^3/3 = (\sigma_0 - \Delta\sigma)r_0^3/3 = \sigma_0 r_0^3/3 - G_0 r_0^3 z/(3\,z_0) \tag{64}$$

From this we find

$$B = G_0 r_0^3/(3\,z_0)$$

The recurrent deformation due to molecular dislocations is given by

$$\gamma_\infty = \frac{\alpha r_0^3 \sigma_0}{6\,kT + \alpha G_0 r_0^3} \tag{65}$$

(γ_∞ means the reduction of the purely elastic deformation). As γ_0 remains constant, we observe a residual elastic deformation[29)]

$$\gamma_{res} = \gamma_0 - \gamma_\infty = \frac{\sigma_0}{G_0} - \frac{\sigma_0}{G_0 + 6\,kT/(\alpha r_0^3)}$$

and therefore a residual stress

$$\sigma_{res} = \sigma_0 - \frac{\sigma_0}{1 + 6\,kT/(\alpha r_0^3 G_0)} \tag{66}$$

In Fig. 24(a) the purely elastic deformation and the plastic elastic flow processes are plotted and hatched in a different manner. Figure 24(b) shows the dependence of stress on time. It can also be seen, that with discharge at time t_0 the purely elastic residual deformation disappears at once, whereas the plastic-elastic portion does so gradually (diffusion processes).

If there are irreversible flow processes at a given constant deformation γ_0 with $A \ll kT$, we get

$$\frac{dz}{dt} = \frac{\alpha z_0}{h}\left(\frac{r_0^3 \sigma_0}{3} - Bz\right)\exp - \frac{\Delta U}{kT} \tag{67}$$

The solution here is $z = z_\infty\left(1 - \exp - \dfrac{t}{\tau_{irr}}\right)$ with

$$z_\infty = r_0^3 \sigma_0/(3\,B) = z_0 \sigma_0/G_0 \text{ and } \tau_{irr} = \frac{h \exp \Delta U/kT}{\alpha z_0 B}$$

This implies $\gamma_\infty = \sigma_0/G_0$. The total deformation becomes viscous. As in (63) the denominator is greater than in the equation for τ_{irr}

$$\tau_{irr} > \tau_{stress}$$

The stress decreases to zero. After these dislocations the flowing processes will come to an end. The total strain γ_0 remains constant. In Fig. 25 the course of the strain

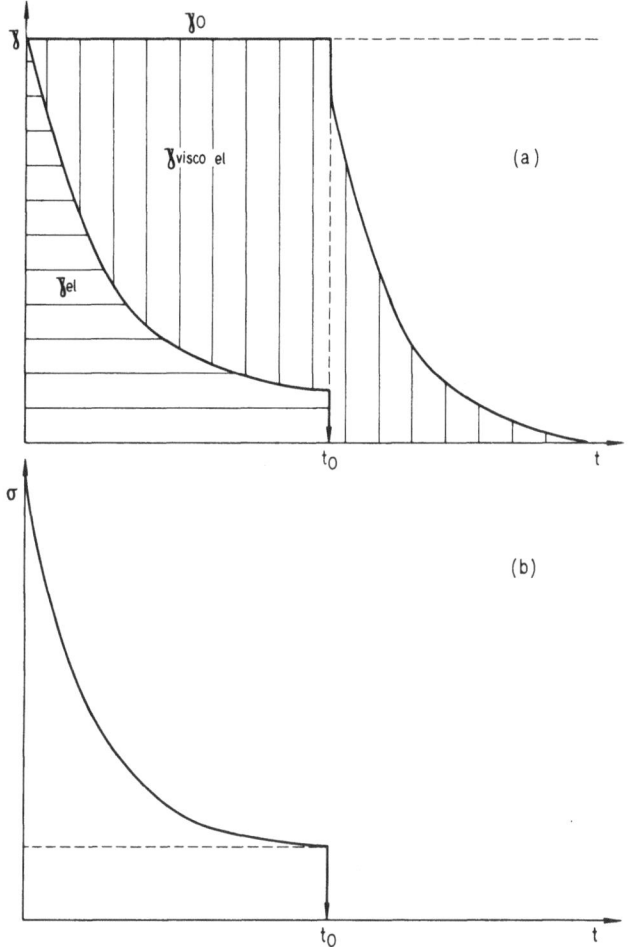

Fig. 24. Elastic and viscoelastic deformation for a constant total deformation γ_0, coupled with a decrease of stress (t_0 unburdening time) (a) $\gamma(t)$, (b) $\sigma(t)$

as a function of time is plotted for the given approximation. As reversible and irreversible flow processes occur simultaneously we observe a predominance of the irreversible dislocations and the disappearance of reversible flow (Fig. 25). We obtain the curve plotted in Fig. 25 as the result of the following system of differential equations describing reversible and irreversible flow jointly:

$$\frac{dz_1}{dt} = \frac{1}{h}\left\{\alpha z_0\left[\frac{r_0^3 \sigma_0}{3} - B(z_1 + z_2)\right] - 2\,z_1 kT\right\}\exp - \frac{\Delta U}{kT}$$

(rev. flow)

$$\frac{dz_2}{dt} = \frac{\alpha z_0}{h}\left[\frac{r_0^3 \sigma_0}{3} - B(z_1 + z_2)\right]\exp - \frac{\Delta U}{kT}$$

(irrev. flow)

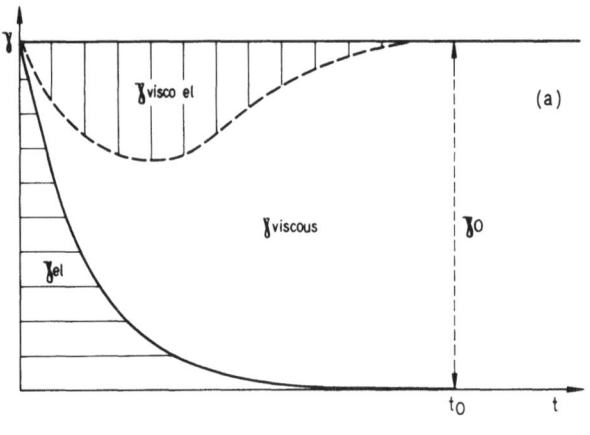

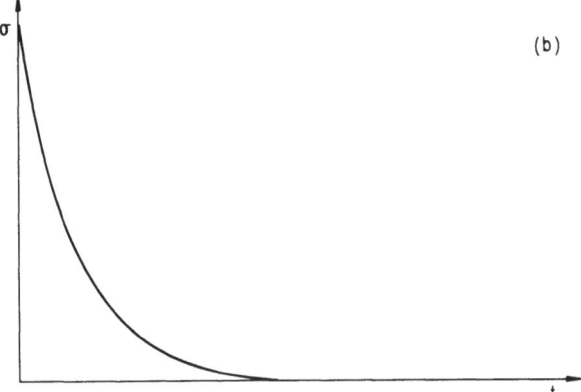

Fig. 25. Distribution of purely elastic, viscoelastic and viscous deformations correlated to a decrease of stress at a constant deformation rate γ_0, with all deformation processes coupled. (a) $\gamma(t)$, (b) $\sigma(t)$

Using the Laplace transformation, we obtain the result

$$z_1 = \frac{\alpha z_0 r_0^3}{\sqrt{G_0^2 \alpha^2 r_0^6/9 + (kT)^2}} \exp\left(-\frac{t G_0 r_0^3 \exp - \Delta U/kT}{3h}\right)$$

$$\cdot \sinh\left(\sqrt{G_0^2 \alpha^2 r_0^6/9 + (kT)^2} \; \frac{t}{h} \exp - \frac{\Delta U}{kT}\right)$$

A similar equation is valid for z_2. The reversible flow processes reach a maximum and then disappear completely.

We should mention that the visible total deformation remains constant (Fig. 25).

In the case of discharge at a specific moment $t < t_0$ we observe residual stresses leading to a recurrent deformation ($\gamma_\infty < \gamma_0$). This occurs primarily when the material is heated.

8.2. Time Dependent Elastic and Electric Potential

Introducing $A = \hat{A}\,e^{j\omega t} = \hat{\sigma}\,r_0^3/3 \cdot e^{j\omega t}$ into the fundamental differential equation, we now have to solve

$$\frac{dz}{dt} = v_{th}[\alpha z_0 \sinh(\hat{A}\,e^{j\omega t}/kT) - 2\,z \cosh(\hat{A}\,e^{j\omega t}/kT)]\exp - \frac{\Delta U}{kT} \qquad (68)$$

Only if $A \ll kT$ is the calculation simple and leads to

$$\frac{dz}{dt} = v_{th}\left[\frac{\alpha z_0}{kT}\,\hat{A}\,e^{j\omega t} - 2\,z\right]\exp - \Delta U/kT \qquad (69)$$

The well known solution yields

$$z = \hat{z}\,e^{j\omega t}/(1 + j\omega\tau) \qquad (70)$$

(the sign $\wedge$ denotes the amplitude).

We then find $\hat{z} = \alpha z_0 r_0^3 \hat{\sigma}/(6\,kT)$ and $\tau = \dfrac{1}{2\,v_{th}}\exp\dfrac{\Delta U}{kT}$. With $\gamma_i = \hat{z}_i/z_0$; α_i and ΔU_i ($i = 1, 2, 3$, related to α, β, γ Process) we obtain

$$\gamma = \sum_i \frac{\alpha_i \cdot \hat{\sigma} \cdot r_0^3}{6\,kT} \cdot \frac{1}{1 + j\omega\tau_i}\,e^{j\omega t} \qquad (71)$$

Using the values

$$\Delta U_i = 1, 3, 5, 10 \text{ and } 20 \text{ kcal/gmol}$$

we get $\tau = 2.6\ 10^{-13}$, $7.4\ 10^{-12}$, $1\ 10^{-10}$, $8.5\ 10^{-6}$ and 15 sec respectively.

As the relaxation times for mechanical losses are about 10^{-6} sec or greater, it is evident that the flow units, which will change their places, are connected with their neighbors by more than one chemical bond. Some chemical binding forces have to be overcome simultaneously otherwise coupled flip-flop-processes will play a dominant role. In phase with the external stress $\sigma = \hat{\sigma}\exp j\omega t$ there is a purely elastic deformation so that

$$\gamma_{el} = \frac{\hat{\sigma}\exp j\omega t}{G_0} \qquad (72)$$

is valid. In the Gaussian plane we find the circle C for the complex shearing rate (Fig. 26) according to Cole Cole[56]. The loss angle is represented by δ. In the case that $\omega\tau_i = 1$, the imaginary component agrees with the real component. Every relaxation process is influenced by the number of participants (α_i). It may be e.g. that nearly all CH_3

groups are mobile but few particles can form kinks. Furthermore we can calculate the real quotient $\hat{\gamma}/\hat{\sigma} = 1/G_0$, whereas

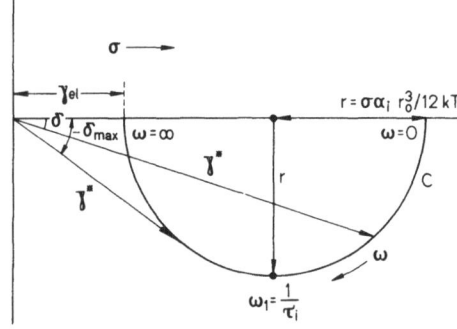

Fig. 26. Arc C for complex deformation γ^* according to Cole and Cole[56] created by periodically changing stress σ (δ is the angle designating the retardation of deformation, γ^* is the sum of the elastic, and viscoelastic deformations)

$$\frac{1}{G_i'} = \frac{\alpha_i r_0^3}{6\,kT} \cdot \frac{1}{1 + \omega^2 \tau_i^2} \quad \text{and} \quad \frac{1}{G_i''} = \frac{\alpha_i r_0^3}{6\,kT} \cdot \frac{\omega \tau_i}{1 + \omega^2 \tau_i^2} \tag{73}$$

are real and imaginary components for the reciprocal shear modules, with

$$\tan \delta_i = \frac{G_i''}{G_i'} \quad \text{and} \quad \tau_i = \frac{h}{2\,kT} \exp \Delta U/kT \tag{74}$$

For all relaxation processes, including the purely elastic deformation, we get

$$\frac{1}{G'} = \frac{1}{G_0} + \sum_i \frac{\alpha_i r_0^3}{6\,kT} \cdot \frac{1}{1 + \omega^2 \tau_i^2} \; ; \quad \frac{1}{G''} = \sum_i \frac{\alpha_i r_0^3}{6\,kT} \cdot \frac{\omega \tau_i}{1 + \omega^2 \tau_i^2} \tag{75}$$

8.3. Dielectric Losses

In conformity with the mechanical losses we can calculate the orientation polarisation using Eq. (28). It will be $A = \frac{1}{3} \mu_0 E_{or} \exp j\omega t$. If there are three possible conformations due to rotation of the dipole moment, using $3\sqrt{2} - \sqrt{6}/3\,\pi \approx 0.19$ we get

$$A = 0.19\,\mu_0 E_{or} \exp j\omega t$$

Analogous to Eq. (70) the number of orientated dipole moments is

$$z = z/(1 + j\omega\tau) \cdot \exp j\omega t \tag{76}$$

with

$$\hat{z} = \frac{\alpha z_0}{2\,kT} \hat{A}_{diel}$$

and

$$\tau = \frac{1}{2\,v_{th}}\exp \Delta U/kT$$

(z, z_0 are related here to the spatial unit).

The z dipoles, that are partially orientated in the direction of the orientation field, supply a dipole moment

$$\hat{\mu}_{or} = \frac{\hat{z}\mu_0/z_0}{3}$$

With Eq. (76) we get

$$\hat{\mu}_{or} = \frac{\alpha \hat{A}_{diel}\,\mu_0}{6\,kT(1+j\omega\tau)} \qquad (77)$$

and with Eq. (28) this becomes

$$\mu_{or} = 0.166\ \alpha\mu_0^2 E_{or}\ \frac{1}{3\,kT(1+j\omega\tau)} \qquad (77a)$$

This equation is true for a stochastic distribution of two possible positions. If there are three conformations containing an angle of 120° we have

$$\mu_{or} \approx 0.095\ \alpha\mu_0^2 E_{or}\ \frac{1}{3\,kT\,(1+j\omega\tau)} \qquad (77b)$$

The polarisation is not in phase with the external electrical field E nor with the orientation field E_{or}.

In the case of two possible positions distributed stochastically for the orientation polarisation we find

$$P_{or} = \frac{z_0\,\mu_0^2\hat{E}_{or}\,\alpha}{18\,kT(1+j\omega\tau)}\exp j\omega t \qquad (78)$$

If a dipole has three possible conformations separated by energy barriers and rotation angles of 120°, we use

$$P_{or} = \frac{0.095\ z_0\,\mu_0^2\hat{E}_{or}\,\alpha}{3\,kT(1+j\omega\tau)}\exp j\omega t \qquad (79)$$

The share rate α of dipoles taking part in forming different conformations increases with the temperature.

Besides the orientation polarisation we observe a deformation polarisation, identical to that of an induced dipole moment. In a spatial unit the polarisation caused by deformation (due to a shift of the electrical charges) will be

$$P_{def} = \frac{\epsilon_\infty - 1}{\epsilon_\infty + 2} \cdot \frac{3 M \epsilon_0}{\rho N_L} \, z_0 \, \hat{E}_{or} \exp j\omega t \tag{80}$$

with ϵ_∞ the dielectric constant for very high frequency, ϵ_0 the dielectric constant of vacuum equal to 8.85416 10^{-12} As/Vm, M the average mass of N_L dipole groups (N_L Avogadro number) and ρ the density.

As in the solid state the share rate α of dipoles taking part in dipole orientation is much more smaller than in the liquid state, where orientation polarisation Eqs. (78) and (79) is also low. The dielectric constant even in polar polymers is not higher than 6.

The average dipole moment for molecular orientation according to Debye

$$\frac{\mu_0^2 \, E_{or}}{3 \, kT} \tag{81}$$

exceeds the polarisation effects is solids. The total polarisation will be $P_{or} + P_{def}$.

9. Coupled Dislocation Processes

In many cases molecular displacements depend on other transition processes occurring previously in the neighborhood. Let us consider Fig. 27. The double kink may move from 1 to 1'. Then segment 2 has the possibility to go from 2 to 2'. Flowing units 4, 5 and 6 may undergo similar processes. At last the molecule may reach the dashed structure. In every case the final position will be reached after some successive place exchange processes each determining the next step. We therefore get a system of simultaneous differential equations.

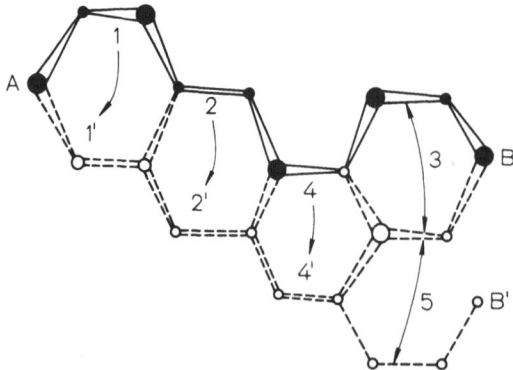

Fig. 27. Coupled molecular dislocation processes (1, 2, 3, 4, 5) which lead to the conformation AB' from the initial conformation AB (out of plane positions)

$$\frac{dz_1}{dt} = \nu_{th}[(z_0\alpha/n - z_1)\exp{-(\Delta U - A)/kT} - (z_0\alpha/n + z_1 - z_2)\exp{-(\Delta U + A)/kT}]$$

$$\frac{dz_2}{dt} = \nu_{th}[(z_0\alpha/n + z_1 - z_2)\exp{-(\Delta U - A)/kT} - (z_0\alpha/n + z_2 - z_3)\exp{-(\Delta U + A)/kT}]$$

$$\frac{dz_3}{dt} = \nu_{th}[(z_0\alpha/n + z_2 - z_3)\exp{-(\Delta U - A)/kT} - (z_0\alpha/n + z_3 - z_4)\exp{--(\Delta U + A)/kT}]$$

$$\vdots \qquad \vdots \qquad\qquad \vdots \qquad\qquad (82)$$

Initially we begin our calculations with n equally probable positions for which $z_0\alpha/n$ is the average population of every conformation possible. The probability of molecular barrier processes occurring in the direction of an external stress increases under the influence of the external stress.

The probability per unit of time is $\nu_{th}\exp - \dfrac{(\Delta U - A)}{kT}$ in the direction of the stress and $\nu_{th}\exp - \dfrac{\Delta U + A}{kT}$ opposite to it.

Using the Laplace-transformation

$$L(J) = \int_0^\infty J(t)e^{-pt}\,dt$$

we obtain a system of n algebraic equations with n unknown functions. The solution of this system can be found in the transformed space of the Laplace-transformation using the method of solving n algebraic equations with n unknown quantities. Transforming this result into the system with the coordinates z and t we get a sum of exponential functions with different relaxation times for the number of place exchange processes.

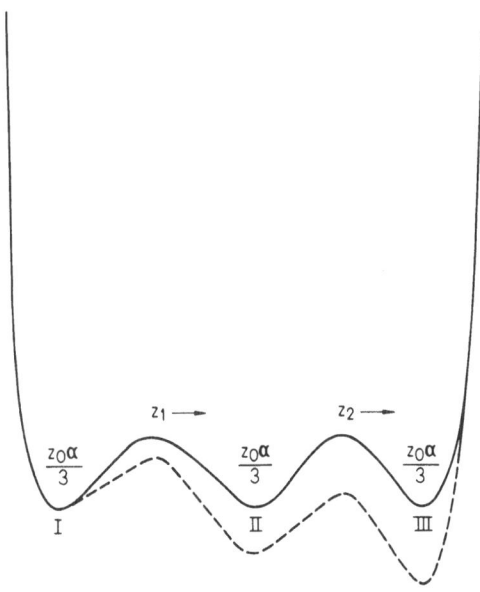

Fig. 28. Molecular dislocations z_1 and z_2 distributed over three possible conformations I, II, III

To study the method we restrict ourselves to the case of three possible positions (Fig. 28) so that

$$\frac{dz_1}{dt} = v_{th}\left[\left(\frac{\alpha z_0}{3} - z_1\right)\exp - \frac{\Delta U - A}{kT} - \left(\frac{z_0\alpha}{3} + z_1 - z_2\right)\exp - \frac{\Delta U + A}{kT}\right]$$

$$\frac{dz_2}{dt} = v_{th}\left[\left(\frac{\alpha z_0}{3} + z_1 - z_2\right)\exp - \frac{\Delta U - A}{kT} - \left(\frac{z_0\alpha}{3} + z_2\right)\exp - \frac{\Delta U + A}{kT}\right], \quad (83)$$

where z_1 and z_2 are the number of segments going from positions 1 to 2 and 2 to 3 respectively. The total number of molecular dislocations is $z = z_1 + z_2$. It is sufficient to calculate only z. For the shear rate z/z_0 we get

$$\gamma = \frac{2\,v_{th}\alpha}{3}\,\sinh\frac{A}{kT}\left\{\frac{1 + \cosh\dfrac{A}{kT}}{2\cosh\dfrac{A}{kT} - 1}\left(1 - \exp - \frac{t}{\tau_1}\right)\right.$$

$$+ \frac{1 - \cosh\dfrac{A}{kT}}{2\cosh\dfrac{A}{kT} + 1}\left.\left(1 - \exp - \frac{t}{\tau_2}\right)\right\}$$

with

$$\tau_1 = \frac{\exp \Delta U/kT}{v_{th}(2\cosh A/kT - 1)} \qquad \text{and}$$

$$\tau_2 = \frac{\exp \Delta U/kT}{v_{th}(2\cosh A/kT + 1)} \qquad\qquad (84)$$

This solution is characterized by two different relaxation times, whereas the fundamental relaxation time is two times longer than in the case of a single barrier process.

The main effect for coupled molecular displacements is that the relaxation processes are prolonged.

The calculation of the relaxation phenomena was carried out in Leipzig with the assistance of a computer.

In the case of 10 coupled dislocation processes, we obtained a value for the main relaxation time $\bar{\tau} = 2.3 \cdot 10^{-8}$ s using $\Delta U = 10\,kT$ and $A = 0.1\,kT$ whereas the value of $\bar{\tau}$ was about $0.4 \cdot 10^{-8}$ s for 3 coupled processes ($n = 3$) (Fig. 29).

For the case $n = 3$ we initially assumed $z_0/3$ realized conformations for every possibility and set

$$\sinh\frac{A}{kT} \approx \frac{A}{kT};\; \cosh\frac{A}{kT} \approx 1 \text{ and } 1 - \cosh\frac{A}{kT} = -\frac{1}{2}\left(\frac{A}{kT}\right)^2;\; \frac{v_{th}A}{kT} = \frac{A}{h}$$

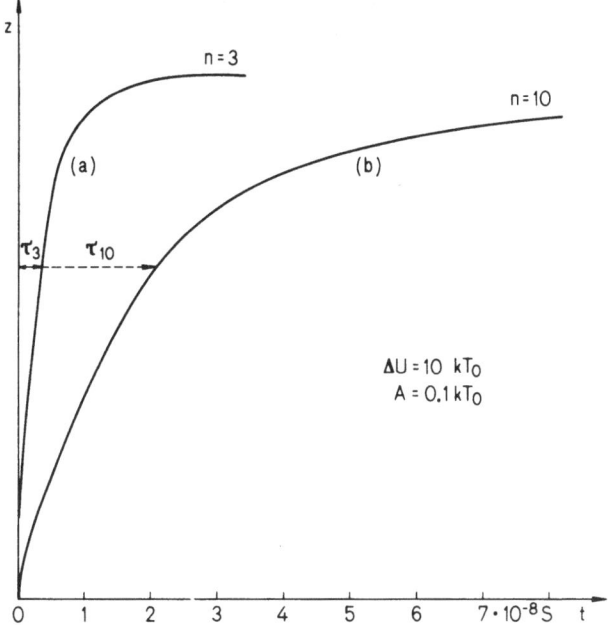

Fig. 29. Number z of molecular dislocations for coupled exchange processes
z as a function of 3 coupled conformations (curvature a)
z as a function of 10 coupled conformations (curvature b)

We found

$$\gamma = \frac{2}{3} \alpha \left[\frac{2A}{h} (1 - \exp - t/\tau_1) - \frac{A^3}{6 h(kT)^2} (1 - \exp - t/\tau_2) \right]$$

with

$$\tau_1 = \frac{1}{\nu_{th}} \exp \Delta U/kT \text{ and } \tau_2 = \frac{1}{3 \nu_{th}} \exp \Delta U/kT$$

10. Rubberlike Behavior, from the Point of View of the Theory of Molecular Displacements

The caoutchouc elastic processes are characterized by many possible molecular conformations separated by very-low energy barriers ΔU. The existence of cross-links
that is "understructible" bonds between the molecules, causes the termination of
the flow processes after orientation. The network can be illustrated by potential
functions like those in Fig. 28. The relaxation processes do not play any decisive
role in ideal caoutchouc. The low activation energy leads to very short relaxation
times. The distribution of the possible molecular conformations will be changed

under the influence of an external stress. The potential functions are superposed on an elastical potential A such that the neighboring conformations leading to an orientated molecule differ by the energy A.

Since we are interested only in the equilibrium conditions $\dfrac{dz_i}{dt} = 0$ and $\Delta U \approx 0$ instead of Eq. (82) for the number of molecular dislocations over small barriers z_1, z_2 ..... we find

$$\frac{z_0\alpha}{n} \sinh \frac{A}{kT} - z_1 \cosh \frac{A}{kT} \qquad\qquad + \frac{z_2}{2} \exp - \frac{A}{kT} = 0$$

$$\frac{z_0\alpha}{n} \sinh \frac{A}{kT} - z_2 \cosh \frac{A}{kT} + \frac{z_1}{2} \exp \frac{A}{kT} + \frac{z_3}{2} \exp - \frac{A}{kT} = 0$$

$$\frac{z_0\alpha}{n} \sin \frac{A}{kT} - z_3 \cosh \frac{A}{kT} + \frac{z_2}{2} \exp \frac{A}{kT} + \frac{z_4}{2} \exp - \frac{A}{kT} = 0$$

$$\vdots \qquad\qquad \vdots \qquad\qquad \vdots \qquad\qquad \vdots$$

$$\frac{z_0\alpha}{n} \sin \frac{A}{kT} - z_n \cosh \frac{A}{kT} + \frac{z_{n-1}}{2} \exp \frac{A}{kT} \qquad\qquad = 0$$

and after summation

$$\frac{z_0\alpha}{n} \cdot n \sinh \frac{A}{kT} = \frac{z_1}{2} \exp - \frac{A}{kT} + \frac{z_n}{2} \exp \frac{A}{kT}$$

The exact solution of this system is complicated. In a rough approximation we assume the quantity for all dislocations from one conformation to the neighboring conformation to be equal ($z_1 = z_2 = \ldots = z_n$). In reality $z_n > z_1$. This assumption leads to $z = n\,z_1$.

Therefore the function which gives the number of exchange processes before saturation occurs will be

$$z = n \cdot z_0 \cdot \alpha \cdot \tan h\, A/kT$$

or

$$\gamma = h \cdot \alpha \cdot \tanh \sigma r_0^3/3\, kT, \text{ resp. } \gamma = h\alpha \tanh(\text{const } \sigma^2/kT) \text{ (with Eq. (21)) and } \sigma_2 = 0$$

Introducing the ratio of the stretched length to unstretched length

$$\Lambda = \frac{L + \Delta L}{L} = 1 + \gamma^{*)}$$

---

*) In this chapter γ denotes the elongation.

we get

$$\frac{\Lambda - 1}{n\alpha} = \tanh \sigma r_0^3 / 3\,kT \quad \text{resp.} = \tanh \text{const } \sigma^2 / kT$$

Calculating σ as a function of Λ leads to

$$\sigma = \frac{3\,kT}{r_0^3} \cdot \text{Ar} \tan \frac{\Lambda - 1}{n\alpha} \quad \text{resp. } \sigma = \sqrt{\text{const Ar} \tan \frac{\Lambda - 1}{n\alpha}}$$

The number of possible conformations n for each mobile molecular segment is reciprocal to the density of cross-links N (defined for 100 C–C groups in the polymer chain).

The deformation γ or Λ decreases with increasing density of the cross-links (*e.g.*) an increase of the sulphur bridges in vulcanized caoutchouc). Moreover the saturation stress σ_{max} increases with the temperature. A better reproduction of the experimental values can be obtained by using $A = \text{const } \sigma^2$ in the molecular exchange theory or with the empirical Mooney-Rivling equation

$$\sigma = (C_1 + C_2/\Lambda)(\Lambda - \Lambda^{-2})$$

This equation shows discrepancies between experiment and theory in the saturation range. It gives better results for low stresses as compared with our approximation equation obtained from a simplified quasicubic model for molecular dislocations (Fig. 30)[57].

It should be added that the considered system is only truly for ideal caoutchouc. In reality the deformation processes are superposed on slowly occurring molecular dislocations. We observe new conformations in the direction of an external stress

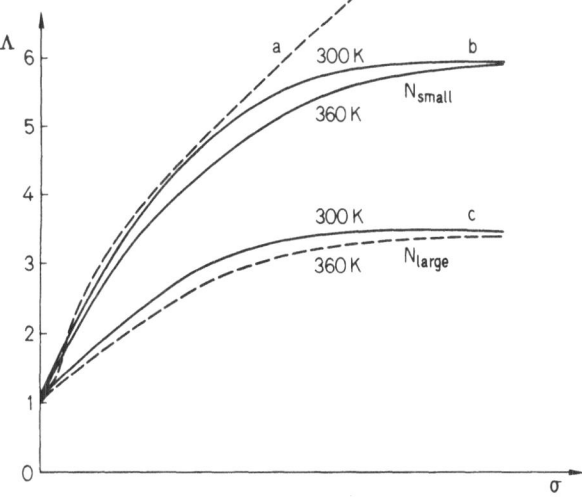

Fig. 30. Ideal caoutchouc elastic deformation leading to an elongation Λ as a function of stress (a) Mooney-Rivling equation, (b) and (c) place exchange theory

characterized by high activation energies primarily in the case of high elastic synthetic materials[47].

These processes can be described by the theory presented here [Eqs. (30) to (60)], but in general they consist of coupled processes with very long relaxation times.

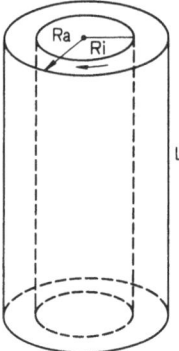

Fig. 31. Couette voscosimeter

To a first approximation, the maximum elongation due to stretching effects can be calculated using the statistical length L of the molecules according to Kuhn, $L = \sqrt{n} \cdot a$, (n is the number of segments free to move and a the length of a monomer unit. With $n \cdot a$ as the maximum length we obtain

$$\Lambda = \frac{n \cdot a}{\sqrt{n} \cdot a} = \sqrt{n}$$

Assuming $n = 60$ (about 1,5% cross-links) we get $\Lambda \approx 8$ corresponding to an elongation of about 800%.

Conclusions

A simple model of quasicubic elements of flow traversing barriers determined by chemical bonds describes the mechanical and dielectric relaxation and viscoelastic as well as viscous behavior. Though we consider only one dimensional molecular displacements related to the direction of an external stress or electric field, experiments and theory are in good agreement. For both the dependence on temperature and the existence of several relaxation ranges a good explanation can be found even in the case of nonlinear behavior. The assumption of monochromatic thermal vibrations (10^{13} Hz) is a very rough approximation. The stimulation of different thermal vibrations, taking into account the moments of inertia of the molecular segments, broadens the spectrum of the relaxation times. This was calculated by Das[58]. Furthermore the specific morphology of molecular chains and their mobility supplies details of the dislocation processes. Pechhold[59, 60] and co-workers have done extensive work here. For our part we think it may be useful to continue the investigations of coupled dislocation processes particularly in the case of caoutchouc.

Appendix

I. The theory of molecular dislocations used to describe deformation and relaxation is based on the assumption of a distribution of the thermal vibration energy similar to that applicable to gas molecules. In general we consider the superposition of thermal motion in one direction to be given by the geometric position of two possible conformations. In this single dimension the phase space elements are dx (space coordinates) and dp (momentum coordinates). The sum of states

$$\int_0^\infty \exp - \frac{-p_x^2}{2\,mkT} \cdot dp_x = \sqrt{\pi mkT/2}$$

is the same as in the case of gas molecules. For the probability W_p of finding a particle which vibrates symmetrically to the minimum of the energy hyperplane with a pulse rate between p_x and $p_x + dp_x$ we obtain

$$W_p \cdot dp_x = \frac{\exp - \dfrac{p_x^2}{2\,mkT}\,dp_x}{\sqrt{\pi mkT/2}}$$

in good agreement with Ref.[8].

Substituting $U_1 = p_x^2/2\,m$ and $dp_x = \sqrt{\dfrac{m}{2\,U_1}}\,dU_1$, where $U_1 > 0$ we get

$$W_U dU_1 = \frac{1}{\sqrt{\pi U_1 kT}}\,\exp - \frac{U_1}{kT}\,dU_1$$

The total kinetic energy U contributing to molecular dislocations in the given direction $A \longleftrightarrow C$ (one dimensional consideration) may be supplied by more than one thermal motion simultaneously (e.g. oscillations and hindered rotations). In all cases where the total energy U is given by $U_1 + U_2 = U$ the probability for U will be calculated by multiplication and integration over all cases satisfying the condition $U_2 = U - U_1$,

$$W_U dU = \frac{dU}{\pi kT} \int_{U_1 \to 0}^{U} \frac{1}{\sqrt{(U - U_1)U}}\,\exp - \frac{(U - U_1)}{kT}\,\exp - \frac{U_1}{kT}\,dU_1$$

(the probability W_U is equal to the sum of states satisfying the condition $U_1 + U_2 = U$).
The result of the integration is

$$W_U dU = \frac{1}{\pi kT}\,\text{arc sin}\left(\frac{2\,U_1 - U}{U}\right)_{U_1=0}^{U_1=U}\,\exp\left(-\frac{U}{kT}\right)dU$$

so that

$$W_U dU = \frac{\exp - U/kT}{kT}\,dU$$

We can now calculate the probability that $U > \Delta U$. We find

$$W_p = \int_{\Delta U}^{\infty} \frac{\exp - U/kT}{kT}\, dU = \exp - \frac{\Delta U}{kT}$$

The Arrhenius formula is the result of the Boltzmann statistics based on the super-position of two thermal vibrations. For four oszillations, of which two have the same share rate U_1, whereas the other two have the share rate, $U - U_1$ we find

$$W_U dU = \frac{dU}{(kT)^2} \int_0^U \exp - (U - U_1) \exp - U_1\, dU_1 = \frac{\exp - U/kT \cdot U}{(kT)^2}$$

In this case the probability of overcoming the potential barrier ΔU is:

$$W_p = (1 + \Delta U/kT)\exp - \Delta U/kT$$

Aussuming that only a share rate β $(0 < \beta < 1)$ of the thermal vibrational energy of the particles oscillating at a greater distance contributes to the molecular dislocations for the probability considered of obtaining the efficient energy U, we get

$$W_U dU = dU \int_0^{U/\beta} \frac{\exp - (U - U_2\beta)/kT}{kT} \cdot \frac{\exp - U_2/kT}{kT}\, dU_2$$

It should be noted that $U_1 = U - \beta U_2$ and U_2 are limited by U/β. After integration we find

$$W_U dU = \frac{\exp - U/\beta kT - \exp - U/kT}{kT(\beta - 1)}\, dU$$

The probability for molecular dislocations is calculated by integration between the limits ΔU_0 and $U = \infty$. This yields

$$W_p = \frac{\beta \exp - \Delta U/\beta kT - \exp - \Delta U/kT}{\beta - 1}$$

Finally we must consider the case that n thermal vibrations are participating only with the share rate U/n. This is quite the same as if the chemical binding energy ΔU were much greater, in our case n times greater.

Since the thermal vibration energy is only participating in overcoming a given barrier described by the activation energy ΔU with the share rate 1/2, 1/3, 1/4 etc, we obtain results that are 2, 3, 4 etc times higher than we had used the Arrhenius equation in the traditional manner.

The probability considered is valid for only one thermal vibration. Since to a first approximation there are $\nu_{th} \approx 10^{13}$ vibrations, the probability of overcoming a barrier ΔU given by chemical forces is ν_{th} times greater. With the Planck constant h we find $h \nu_{th} = kT$ and $\nu_{th} = kT/h$.

II. *Viscous flow through a capillary and flow processes between rotating cylinders.*
We shall try to calculate the viscous behavior for non-Newtonian liquids flowing through a capillary. Beginning with $\eta = \sigma / \dfrac{dv}{dR}$ and

$$\sigma = \frac{\Delta p \cdot R^2 \pi}{2 \pi L R}$$

as well as $\dfrac{dv}{dR} = \dfrac{d\gamma}{dt}$ and $A = \Delta p \cdot R \cdot r_0^3/6 L$ we get

$$\frac{dv}{dR} = \alpha \nu_{th} \sinh \Delta p \, R \, r_0^3/(6 \, LkT) \cdot \exp - \Delta U/kT$$

Here R is the distance from the centre of the capillary, Δp is the difference of pressure over the length L of the capillary and R_a is the radius. Integration yields

$$v = \frac{6 \, \alpha \nu_{th} \, LkT}{\Delta p \cdot r_0^3} \left(\cosh \frac{\Delta p \, R \, r_0^3}{6 \, LkT} - \cosh \frac{\Delta p \, R_a r_0^3}{6 \, LkT} \right) \exp - \frac{\Delta U}{kT}$$

$v = 0$ for $R = R_a$ $(R < R_a, v < 0)$.

For the dependence on R we get a cosh function instead of a parabolic function as is the case for Newtonian liquids. Likewise, it is easy to calculate the volume flowing through the capillary in a unit of time.
After integration we get

$$G = \frac{12 \pi \, LkT \, \alpha \nu_{th}}{\Delta p \cdot r_0^3} \exp - \frac{\Delta U}{kT} \left[\frac{R_a^2}{2} \cosh \frac{\Delta p \cdot R_a \cdot r_0^3}{6 \, LkT} \right.$$
$$\left. - \frac{6 \, LkT \, R_a}{\Delta p \cdot r_0^3} \sinh \frac{\Delta p \cdot R_a r_0^3}{6 \, LkT} + \left(\frac{6 \, LkT}{\Delta p \, r_0^3} \right)^2 \left(\cosh \frac{\Delta p \, r_0^3 R_a}{6 \, LkT} - 1 \right) \right]$$

The flow rate

$$G = \frac{\pi \cdot \Delta p \cdot R_a^4}{8 \eta \, 1}$$

according to Hagen-Poiseuille is less than the flow rate for a non-Newtonian liquid using the place exchange theory.
It is also possible to calculate the flow path in a Couette apparatus for non-Newtonian liquids (flowing between rotating cylinders Fig. 31) (see page 56). If an inner cylinder rotates at an angular velocity ω and a shear deformation takes place in the gap between the internal and external cylinder $(R_i - R_a)$ we observe a torque M:

$$M = \sigma \cdot 2 \pi \cdot R_i^2 \cdot L$$

(L length of the cylinder; R_i and R_a inner and outer cylinder (radius) respectively.)
The viscosity for the Couette flow is given by

$$\eta = \frac{\sigma}{\dfrac{dv}{ds}} = \frac{M[1-(R_i/R_a)^2]}{2\pi L R_i^2 \cdot 2\omega} \cdot \text{with} \quad \frac{dv}{ds} = \frac{2\omega}{1-(R_i/R_a)^2}$$

Setting

$$\frac{dv}{ds} = \frac{2\omega}{1-(R_i/R_a)^2} = \alpha\nu_{th} \; \sinh \frac{A}{kT} \exp - \frac{\Delta U}{kT}$$

with $A = M/(2\pi R_i^2 L)r_0^3/3$ and eliminating M we get

$$\eta = \frac{3(1-(R_i/R_a)^2)kT}{2\omega r_0^3} \text{Arsin} \frac{2\omega \exp \Delta U/kT}{[1-(R_i/R_a)^2]\alpha\nu_{th}}$$

Since Arsin $C \cdot \omega$ increases less than ω, η decreases with

$$\frac{\text{Arsin } C\omega}{\omega} \quad (C = \text{const.})$$

We observe a decreasing viscosity, if the velocity in the measuring device increases.
It is also possible to eliminate ω and to introduce M. In this case we also observe a
decreasing viscosity with increasing M.

References

[1] Stuart, H. A.: Die Physik der Hochpolymeren, Springer-Berlin-Göttingen-Heidelberg 1953,
1956
Saito, S.: Kolloid Zs. *189*, 116 (1963)
Debye, P., Bueche, F.: J. Chem. Phys. *19*, 589 (1951)
[2] Williams, M. J., Landel, R. F., Ferry, I. W.: Amer. Chem. Soc. *77*, 4701 (1955)
Tobolsky: Structure and properties of polymers. New York: Wiley 1966
Kovács, A. J.: Advan. Polymer Sci. *3*, 394 (1964)
McCrum, N. G., Read, B. E., Williams, G.: Anelastic and dielectric effects in polymer solids.
New York: Wiley 1967
Bartenew, G. M., Zugev, Y. S.: Strength and failure of viscoelastic materials, London-New
York: Pergamon Press 1968 (transl. from Russian)
[3] Prandtl, L. Z.: Zs. Angew. Math. u. Mech. *8*, 85 (1928)
[4] Glasstone, S., Laidler, K. I., Eyring, H.: The theory of rate processes New York: McGraw
Hill 1941
[5] Holzmüller, W.: Phys. Zs. *41*, 499 (1940)
Holzmüller, W.: Phys. Zs. *42*, 273 (1941)
Holzmüller, W.: Zs. Phys. Chem. *202*, 330 (1954); *203*, 163 (1954)
[6] Kirkwood, J. G.: J. Chem. Phys. *7*, 911 (1939); *4*, 592 (1936)
[7] Kirkwood, J. G., Fuoss, R. M.: J. Chem. Phys. *9*, 329 (1941)

[8] Holzmüller, W.: Koll. Zs. u. Zs. f. Polym. *203*, 7 (1965)

[9] Holzmüller, W.: Koll. Zs. *155*, 119 (1957)

[10] Holzmüller, W., Altenburg, K.: Physik d. Kunststoffe, Berlin: Akadem. Verl. 1961

[11] Holzmüller, W.: Plaste u. Kautschuk *21*, 329 (1974)

[12] Holzmüller, W., Pitzschke, H., Wendisch, P.: Koll. Zs. u. Zs. f. Polymere *216*, 26 (1967)

[13] Rossini, F. D.: Chemical Thermodynamics. New York: Wiley 1950

[14] Kästner, S.: Koll. Zs. u. Zs. f. Polymere *156*, 142 (1958); *157*, 138 (1958)

Wobser, G.. Hägele, P. C.: Ber. Bunsengesellsch. *74*, 896 (1970)

Van Krevelen, D. W.: Properties of polymers and correlations with chemical structure. Amsterdam-London-New York: Elsevier Publ. Comp. 1972

[15] Schrödinger, E.: Stat. Thermodynamik, Leipzig: J. A. Barth 1952

[16] Baur, H.: Zs. f. Naturforschg. *26a*, 979 (1971)

[17] Zaeschmar, G.: Koll. Zs. and Zs. f. Polymere *177*, 35 (1961); *181*, 7 (1962)

[18] Fowler, R., Guggenheim, E. A.: Statistical thermodynamics. Cambridge: University Press 1956

[19] Großöhme, T., Wünsche, P.: Zs. d. KMU *21*, 644 (1972)

Roth, H.-K.: Dissertation B Leipzig 1976, Wiss. Zs. KMU *21*, 637 (1972)

[20] Hirai, H., Eyring, H.: J. Polym. Science *37*, 51 (1959)

[21] Dugdale, J. S., McDonald, D. K. C.: Phys. Rev. *96*, 57 (1954)

[22] Adam, G.: Koll. Zs. and Zs. f. Polymere *180*, 11 (1962)

[23] Jenckel, E.: Kolloid Zs. *100*, 163 (1942)

[24] Breuer, H., Rehage, G.: Koll. Zs. and Zs. f. Polymere *216–217*, 159 (1967)

[25] Bondi, A.: Physical properties of molecular crystals, liquids and glasses. New York: John Wiley 1968

Marei, A. I.: Kautch. i. Resina. *19*, 2 (1960)

Wunderlich, B., Jones, L. D.: J. Macromol. Sci. *3*, 67 (1969)

[26] Illers, K.-H.: Makromolekulare Chemie. *38*, 168 (1960)

[27] Becker, R.: Zs. Phys. Chemie, Leipzig (to be published)

[28] Gordon, M., Taylor, J. S.: J. Appl. Chem. *2*, 493 (1956)

[29] Boyer, R. F.: Rubber Chem. Technol. *36*, 1303 (1963)

[30] Kanig, G.: Koll. Zs. and Zs. f. Polymere. *233*, 829 (1969)

[31] Kovács, A. J.: Rheol. Acta *5*, 262 (1966)

[32] Onsager, L. J.: Amer. Chem. Soc. *58*, 1486 (1936)

[33] Fröhlich, H.: Theory of Dielectrics, 2. printing. Oxford: University Press 1958

[34] Yamafuji, K.: Koll. Zs. and Zs. f. Polymere. *195*, 111 (1974)

Yamafuji, K.: J. Phys. Soc. Japan *15*, 2295 (1960)

[35] Ishida, Y.: Koll. Zs. *174*, 124 (1961); *174*, 162 (1971)

Ishida, Y., Yamafuji, K.: Koll. Zs. and Zs. f. Polymere. *183*, 15 (1963)

[36] Kirkwood, J. G.: J. Polym. Sci. *12*, 1 (1944); J. Chem. Phys. *14*, 51 (1946)

[37] Saito, S.: Res. Electrotechn. Lab. Tokyo. *1964*, 648

[38] Hoffman, J. D.: 8. Ampere Colloque. *12*, 36 (1959)

Hoffman, J. D.: J. Chem. Phys. *20*, 541 (1952); *22*, 156 (1954); *23*, 1331 (1955)

[39] Gottlib, Yu. Ya., Wolkenstein, N. V.: J. Techn. Phys. *23*, 1936 (1953)

Gottlib, Yu. Ya., Solichow, K. M.: Solid State Phys. *4*, 2461 (1962)

Gottlib, Yu. Ya., Darinskij, A. A.: Wys. mol. soed. *6*, 2938 (1964); *8*, 580 (1967)

[40] Yamafuji, K.: Koll. Zs. and Zs. f. Polym. *195*, 111 (1964)

[41] Williams, G.: Trans. Farad. Soc. *59*, 1397 (1963); *60*, 1548, 1556 (1964); *61*, 1564 (1965)

[42] Slater, J., Kirkwood, J. G.: J. Phys. Rev. *37*, 682 (1931)

Guggenheim, E. A.: Trans. Far. Soc. *37*, 271 (1941)

[43] McGrum, N. G., Read, B. E., Williams, G.: Anelastic and dielectric effects in polymer solids. New York: Wiley 1967

[44] Holzmüller, W.: Koll. Zs. and Zs. f. Polymere. *205*, 24 (1965)

[45] Holzmüller, W., Ilberg, W.: Rheol. Acta *5*, 1 (1966)

[46] Holzmüller, W.: Pure and Appl. Chemistry London. *12*, 359 (1966)

[47] Bartenev, G. M., Selenew, Yu. V.: Material Science and Engineering (Netherland). *2*, 137 (1967)

[48] Holzmüller, W., Jung, P.: Plaste u. Kautschuk. *2*, 218 (1955)
[49] Holzmüller, W.: Plaste u. Kautschuk. *17*, 2 (1970)
[50] Holzmüller, W.: Plaste u. Kautschuk. *20*, 249 (1973)
[51] Müller, F. H., Hellmuth, E.: Koll. Zs. and Zs. f. Polymere. *181*, 97 (1962)
[52] Böttger, H., Bryksin, V. V.: phys. stat. sol. (b). *78*, 9, 415 (1976); *68*, 285 (1975)
     O'Dwyer, J. J.: The theory of electrical conductivity and breakdown in solid dielectrics.
     Oxford 1973
[53] Koppelmann, J., Gieleßen, J.: Koll. Zs. *175*, 97 (1961); Zs. für Elektrotechn. u. Ber. Bunsenges.
     65, 689 (1961)
[54] Kubat, J.: Archiv für Physik. *25*, 285 (1963); *22*, 465 (1962)
[55] Eyring, H., Ree, R.: Proc. Nat. Acad. Sci. USA. *41*, 118 (1955)
[56] Cole, H., Cole, K. S.: J. Chem. Phys. *9*, 341 (1941)
[57] Treloar, L. R. G.: The physics of rubber elasticity. Oxford: University Press 1967
[58] Das, T. P.: Chem. Phys. *25*, 896 (1956), *27*, 763 (1957)
[59] Hägele, P. C., Pechhold, W.: Koll. Zs. u. Zs. f. Polymere *241*, 977 (1970)
[60] Blasenbrey, S., Pechhold, W.: Koll. Zs. u. Zs. f. Polymere *241*, 955 (1970)
     Wobser, G., Hägele, P. C.: Ber. Bunsenges. *74*, 896 (1970)

Received October 15, 1977, November 10, 1977
H.-J. Cantow (editor)

The Iso-Free-Volume State and Glass Transitions in Amorphous Polymers:
New Development of the Theory

Yury Lipatov

Institute of Macromolecular Chemistry, Ukrainian Academy of Sciences, 252160 Kiev, USSR

Table of Contents

1. The Concept of Free-Volume and Its Connection with Glass Transition in Amorphous Polymers

Many important properties of compounds of low and high molecular weight (viscosity and thermal, mechanical and relaxation properties, etc.) are associated with their molecular mobility and the degree of molecular ordering of their structure. The structure and properties of such systems may be described from the thermodynamic point of view by means of such parameters as configurational free energy[1], and enthalpy[2]. A convenient way of describing these properties phenomenologically is by the concept of excess or free-volume. Although much has been published in this field (reviews[3-5]), there are many unresolved problems. According to Covacs[6], free-volume may be considered a characteristic of structure disordering and determines the rate of molecular rearrangement. This characteristic, being closely connected with the configurational mobility of liquids and their thermodynamic functions, could be used as one of the fundamental parameters of the liquid state if one were able to give a satisfactory definition of this concept and find methods of determining the fractional free-volume.

Indeed, some properties of liquids, such as their viscosity and ability to form glasses, point to the existence of a free-volume due to holes of molecular dimensions or voids resulting from irregularities of molecular packing. Lattice models are widely used to describe the properties of polymers[7, 8]. In a lattice model for a low-molecular-weight system the volume is divided into cells of equal size and, if the number of molecules in the system is assumed to be equal to the number of cells, the lattice theory is called free-volume theory[7]; if the number of cells exceeds the number of molecules, it is referred to as hole theory. If the number of cells is equal to the number of the molecules, the size of the cell will always be greater than the inherent volume of the molecule, since the total molecular volume is less than the entire volume of the body even at maximal packing density. In the free-volume theory, when considering the canonic ensembles[9], the principle is:

$$F = -kT \ln Z \tag{1}$$

where

$$Z = Q/(n! \, \Lambda^{3n}) \tag{2}$$

In this case the configuration integral is calculated by assuming that at high densities every cell is occupied by only one molecule because of strong repulsion forces, while for low densities a correction is introduced for multiple occupation of the cell:

$$Q = \eta^n Q^{(1)} \tag{3}$$

where the factor η determining the collective entropy of the system is equal to unity at high, and e at low densities. If ϕ_0 is the potential energy of the lattice in the undisturbed state and $\varphi(R_i)$ the increase in molecular energy when it differs by R_i from the equilibrium state, then

$$Q^{(i)} = \exp(-\phi_0/kT) \, v_f^n \qquad (4)$$

where

$$v_f = \int_\Delta \exp(-\varphi(R)/kT) dR \qquad (5)$$

Since Eq. (5) is equivalent to that of the Einstein model according to which every molecule in the crystal lattice moves freely in the volume v_f of the constant potential ϕ_0, then v_f is a free-volume.

In hole theory the volume of a cell is assumed to be such that at any density the possibility of multiple occupation can be neglected. Moreover, the number of cells being assumed to be greater than the number of molecules, the system can be regarded as consisting of molecules and holes and therefore such a lattice model is called a hole model. In hole theory the zero potential ϕ_0 of the lattice is used as in free-volume theory, but here the increase in molecular energy $\varphi(R_i)$ depends not only on R_i, but also on the number of neighboring molecules and the way they are arranged around the molecule. If the fraction of empty neighboring cells is w, then

$$v_f^{(w)} = \int_\Delta \exp(-\bar{\varphi}(R, w)/kT) \, 4R^2 \, dR \qquad (6)$$

and

$$Q = \exp(-\phi_0/kT) \, \Sigma \exp(nl \, \phi_a/kT) \prod_{i=1}^{n} v_f(w_i) \qquad (7)$$

where ϕ_a is the interaction potential at a distance a between cells and $v_f(w)$ is a generalized free-volume. If all the neighboring cells are occupied ($w = 0$), then $v_f(w)$ becomes equal to the smooth free-volume[7] v_f from Eq. (6), but if some neighboring cells are empty ($w = 1$), then $v_f(w)$ is equal to the volume of the cell q. In Eq. (6) $\bar{\varphi}$ is the number of cell-hole pairs for a given molecule and R is the phase volume. These relationships enable us to determine the limiting form of the configuration integral for high (8) and low (9) densities:

$$Q = v_f^n \exp(-nl \, \phi_a/2 \, kT) \qquad (8)$$

$$Q = q^n l! /(n! \, (l - n)!) \approx \exp(n)(V/n)^n \qquad (9)$$

It follows from what has been said above that free-volume is a value that is determined by both hole volume and empty volume, the latter being connected with the packing mode. In this case an empty volume $v_e = v_r - v_w$, where v_r is the real (observed) volume at temperature T and v_w the volume of the substances as calculated from the Van der Waals dimensions obtained by an X-ray diffraction method or from the gas-kinetic collision cross-section[3]. Then the expansion volume $v_{ex} = v - v_0$, where v_0 is the volume occupied by the molecules at 0 K in a close-packed crystalline state. This expansion volume is extra free space generated by thermal motion. It must always be smaller than the empty volume.

Haward[3] considered a further definition, a fluctuation volume $v_f = N_A V_Q$, where V_Q is the volume swept out by the center of gravity of the molecules as a result of thermal vibration, and N_A the Avogadro number. The last two definitions, as Haward pointed out, had not always been clearly recognized.

The intrinsic volume of a liquid can be estimated if it is assumed to be characterized by the vanishing of the internal pressure P_i[10]. In this case

$$P_i = T\frac{\alpha}{\chi} - P = 0 \tag{10}$$

where $\chi \equiv -(\partial v/\partial p)_T$ or isothermal compressibility and $P_i = (\partial U/\partial v)_T$ or U-specific internal energy. The intrinsic volume is put equal to the volume at 0 K. This value may be calculated by linear extrapolation of the internal pressure as a function of specific volume. This method is useful because it is generally difficult to determine the zero-point volume of polymers in the liquid state. This zero-point volume allows us to estimate the free-volume of polymers[11].

According to Ferry[12] the free-volume per cm of substance, i. e. the fractional free-volume f, is hard to define exactly and should be regarded as merely a useful semi quantitative concept. Specifically, the thermal expansion coefficients of liquids for the most part reflect the increase in fractional free-volume; only a small part is connected with the anharmonic dependence of potential energy or interatomic and intermolecular distances.

The concept of free-volume may also be based on the empirical relation of Doolittle[13] between viscosity η and fractional free-volume:

$$\ln \eta = a + b/f \tag{11}$$

where a and b are constants and

$$f = (v - v_0)/v \tag{12}$$

where v and v_0 are respectively real and occupied volumes of liquid at temperature T. The occupied volume is to be considered as the total volume actually occupied by the molecules as a consequence of the thermal oscillation of all the atoms of each molecule.

The concept of free-volume appeared to be very useful and was applied for the theoretical description of many processes in liquids, including polymeric solutions and melts. Taking the free-volume concept as a basis, theories were developed for the diffusion of low-molecular-weight compounds into polymers[14, 15], thermal conductivity[16], solution and solubility of polymers[17], etc.

The free-volume concept was applied most widely in the theory of viscoelastic properties of polymers developed by Williams, Landel and Ferry (WLF theory), presented in detail in[12]. According to WLF theory, the changes in liquid viscosity with frequency and temperature from glass temperature T_g to T may be plotted on a single master curve by using the reduction factor

$$a_T = \eta_T/\eta_{T_g} \tag{13}$$

From the Doolittle equation one derives

$$\ln a_T = -\frac{(b/2.303\,f_g)(T - T_g)}{f_g/\alpha_f + T - T_g} \tag{14}$$

where f_g is the fractional free-volume at T_g, b is a constant, and α_f is the temperature coefficient of the increase in free-volume (see below). The values f_g found experimentally according to Eq. (14) are the same for many polymers and equal 0.025 ± 0.003.

The theory of viscoelastic properties based on the free-volume concept was developed in[18], the main idea being that some structural elements are displaced when the free-volume fraction exceeds a critical value f_c. The frequency ν of the determining molecular process may be obtained from the equation

$$\nu = \nu_g \exp\left[-N f_c (1/f - 1/f_g)\right] \tag{15}$$

where ν_g is the frequency at T_g. If, as will be considered below,

$$f = f_g + \alpha_f (T - T_g), \tag{16}$$

then

$$\ln (\nu/\nu_g) = \frac{(N f_c/f_g)(T - T_g)}{T - T_g + f_g/f} \tag{17}$$

This equation coincides with Eq. (14).

As early as 1950 Fox and Flory[19] put forward the idea that the glass temperature corresponds to the iso-free-volume state. This hypothesis was developed by other authors[20, 21] and has won wide acceptance. It has been used with minor modifications as the basis for a number of sophisticated theories of the glass-transition phenomenon[22–24].

Simha and Boyer[22] applied the free-volume concept to describe glass transitions in amorphous polymers. By analogy with Eq. (12), it was assumed that the fractional free-volume f_g at glass-transition temperature T_g is

$$f_g = (\nu_g - \nu_0)/\nu_g = 1 - \nu_0/\nu_g \tag{18}$$

where ν_g and ν_0 are the real and occupied volumes at T_g. If we suppose a linear dependence of specific free-volume on temperature, Eq. (16) will obviously be fulfilled. The thermal expansion coefficient α_f in Eq. (16) may be represented as

$$\alpha_f = (df/dT)_p \approx (1/\nu)\left[(d\nu/dT) - (d\nu_0/dT)\right] = \alpha - \alpha_0 \tag{19}$$

Eq. (18) may also be rewritten in another form:

$$f_g = [v_g - v_{0,1}(1 + \alpha_g T_g)]/v_g = [v_g - v_{0,1}(1 + \alpha_g T_g)]/v_{0,1} \tag{20}$$

where $v_{0,1}$ is the occupied volume at absolute zero and α_g is the expansion coefficient of the glassified liquid below T_g. In this definition, f_g represents the difference between the real volume and the volume the liquid should have at T_g if its volume $v_{0,1}$ expanded with the expansion coefficient α_g proper to the glassy state. The real volume at T_g will then be

$$v_g = v_{0,1}(1 + \alpha_1 T_g) \tag{21}$$

and the fractional free-volume:

$$f_g = \frac{v_{0,1}(1 + \alpha_1 T_g) - v_{0,1}(1 + \alpha_g T_g)}{v_{0,1}} \tag{22}$$

From this equation it follows that

$$(\alpha_1 - \alpha_g) T_g = f_g = K_1, \tag{23}$$

K_1 being a constant independent of other material properties (iso-free-volume concept of T_g).

The constant $K_1 = 0.113$ was found by Simha and Boyer[22] for various polymers and was taken as confirmation of the iso-free-volume concept for the glass-transition point. It is worth noting that according to Boyer the individual values for $\Delta\alpha T_g$ range from 0.082 for natural rubber to 0.13 for polyurethane, in which one might anticipate additional molecular processes to be operative[25].

If we suppose then that at T_g the fractional free-volume is constant for all polymers, having defined this volume as

$$K_2 = \frac{v - v_{0,1}}{v} \tag{24}$$

we can show that

$$K_2 = \alpha_1 T_g \tag{25}$$

In reality, we can present the value K_2 in the form

$$K_2 = \frac{v_{0,1}(1 + \alpha_1 T) - v_{0,1}}{v_{0,1}(1 + \alpha_1 T)} = \frac{\alpha_1 T}{1 + \alpha_1 T} \approx \alpha_1 T \tag{26}$$

According to Eq. (23) the ratio between the linearly extrapolated volumes $v_{0,g}$ and $v_{0,1}$ at $T = 0\ K$ is more or less independent of the nature of the polymer. In this case

$$v_{0,g}/v_{0,1} = (1 + \alpha_1 T_g)/(1 + \alpha_g T_g) = 1 + (\alpha_1 - \alpha_g) T_g - \alpha_g(\alpha_1 - \alpha_g) T_g^2 \approx 1 - K_1 \tag{27}$$

The values $K_1 = 0.113$ and $K_2 = 0.164$ have been found for many polymers. The expressions for K_1 and K_2 thus represent the corresponding states in terms of free-volume.

2. Modes of Expression of Fractional Free-Volume at Glass Temperature

The definitions of the free-volume fraction discussed above are dependent on taking the volume as an initial marking-off state, *i.e.* on the manner of calculation. It is already apparent here that there is some uncertainty regarding the physical meaning of free-volume.

Let us therefore discuss some ways of calculating this value upon which the fractional free-volume at T_g depends. In Fig. 1 the solid line ABC represents the actual course of the change in volume of an amorphous polymer when heated through T_g. Segment AB of slope α_g corresponds to the glassy state and segment BC of slope α_l corresponds to the liquid (rubberlike) state Line DE, which is drawn parallel to AB, describes the thermal expansion of a crystalline polymer. (We assume, as usual[1, 25)] that the coefficient of thermal expansion of the occupied volume α_0 as well as that of the crystal, α_c, is identical to α_g.) Point $v_{0,1}$ in Fig. 1 characterizes the state of maximum packing density of molecules of a hypothetical amorphous polymer at absolute zero This is obtained by linear extrapolation of the "liquid" line BC to $T = 0$ K. Extrapolation of the line AB gives the value $v_{0, g}$. In the Simha-Boyer theory the value of the occupied volume v_0' is determined by extrapolation from $v_{0,1}$ to T_g parallel to the line AB (segment $BF = \Delta\alpha\, T_g$).

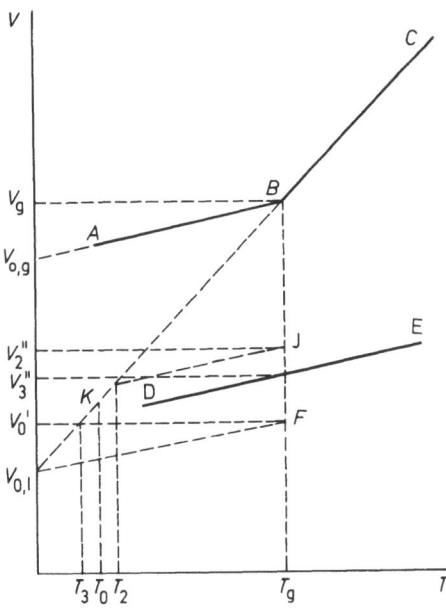

Fig. 1. Schematic representation of the volume-temperature diagram. For definitions, see text

The volume v_2'' in Fig. 1 is the occupied volume in WLF theory. As may be seen from Fig. 1, the value v_2'' is obtained by extrapolation of the liquid volume line at temperature T_2 parallel to AB to T_g (portion BJ). Temperature T_2 is most conveniently determined from the viscosity temperature plot for a supercooled liquid near its T_g[23, 26].

Litt and Tobolsky[27] proposed that the occupied volume v_3'' be defined as that corresponding to the crystalline volume at T_g. It is easily seen from Fig. 1, that all the definitions of the occupied volume are closely related and the differences in definition affect only the numerical value of f_g. According to Cohen and Turnbull[28] and Miller[29], free-volume is determined relative to the temperature T_0 of Vogel's equation describing the temperature dependence of the viscosity[30], point K (Fig. 1) being characterized by the hypothetical volume, which is greater than that for the crystalline state but less than that for glass at the same temperature. Here the segment from point K to the line α_g is equal to the free-volume, which in this case is a little more than the WLF value.

According to Miller[26], $v_f = v - v_0$ at $T = T_0$. Then, at any temperature

$$v_f = \Delta\alpha(T - T_0) \tag{28}$$

where $\Delta\alpha = \alpha_1 - \alpha_g$. Accordingly at T_g

$$v_{fg} = \Delta\alpha(T_g - T_0) \tag{29}$$

According to the Simha-Boyer theory, $\Delta\alpha T_g = K_1$. Equating Eqs. (23) and (29) through $\Delta\alpha$, we can obtain

$$\frac{K_1}{v_{fg}} = \frac{T_g}{T_g - T_0} \tag{30}$$

Difficulties arise, however, concerning the interpretation of T_0 in terms of the free-volume concept. In order to overcome these difficulties, Miller[31] suggested that the occupied volume could vary with temperature.

Covacs[6] proposed another definition of the fractional free-volume. As the extrapolated volume of supercooled liquids reaches the crystalline state, v_c, at temperature $T_c > 0$ K, it is possible to compare this critical temperature with $T_{g\infty}$, the limit glass temperature at infinitely slow cooling. Free-volume in the crystalline state is assumed to be zero. In this case the value f_g according to Doolittle may be compared with $f_{g,c}$, which characterizes the excess free-volume of glass as compared with the crystalline state:

$$f_{g,c} = (v_g - v_c)/v_g \tag{31}$$

To compare f_g and $f_{g,c}$ for amorphous polymers is naturally very awkward, as v_c may be determined experimentally only when the polymer is capable of crystallization. In the case under discussion it is assumed that $\alpha_e \cong \alpha_g$. However, in most cases f_g is twice $f_{g,c}$.

It is generally assumed that fractional free-volume changes with temperature according to Eq. (16), where the thermal expansion coefficient is expressed by Eq. (19).

According to Ferry [12] the macroscopic coefficient of thermal expansion, α_1, above T_g reflects the appearance of the free-volume. In this case the value α_f must be equal to $\alpha_f = \alpha_1 - \alpha_g \cong \Delta\alpha$. Indeed, the experimental values for $\Delta\alpha$ are in good agreement with the theoretically calculated values for α_f.

On the basis of the hole theory of Hirai-Eiring [32, 33] as developed by Smith [34], the volume of the whole system may be given as

$$V = N_0 v_0 + N_h v_h \tag{32}$$

where N_0 is the number of segments of volume v_0 and N_h is the number of holes of volume v_h. Here $V_0 = N_0 v_0$ is the occupied volume and its physical meaning is the specific volume of hypothetical liquid without holes at 0 K. This value may also be used to calculate the fractional free-volume, as can be found experimentally b using the state equation

$$V = V_0 [1 + \sigma^{-1} \exp\left(-(\epsilon_h + PV_h)/kT\right)] \tag{33}$$

where V_h and ϵ_h are the volume and energy of holes, P is the pressure, $\sigma = \exp(1 - n^{-1} - S_h/k)$, and S_h the change in entropy connected with hole appearance, and $n = v_0/V_h$. The fractional free-volume may also be expressed:

$$f = N_h/nN_0 \tag{34}$$

It is clear from the foregoing discussion that there are many ways of expressing fractional free-volume. At the same time, whatever the method used for calculating the free-volume fraction, this value may in all cases be expressed by Eq. (18). As a consequence, the values for f_g calculated according to different theories do not coincide. This circumstance shows again that this value is not a universal physical parameter, but merely a suitable parameter to describe the properties of liquids. As early as 1963 Boyer [25] remarked that free-volume at T_g is constant only under certain conditions and changes with molecular weight, degree of crosslinking, and other characteristics of the polymer.

3. Thermodynamic and Molecular Consideration of the Concept of Free-Volume

Hirai and Eiring [20] established that the value $\Delta\alpha T_g$ is dependent on the energy of hole formation ϵ_h in accordance with the equation:

$$\Delta\alpha T_g = (\epsilon_h/RT_g) \exp - (\epsilon_h/RT_g) \tag{35}$$

Taking $f_g = 0.02$, the theoretical value $\Delta \alpha T_g$ was calculated as 0.08. According to their idea,

$$\epsilon_h / RT_g = -\ln f_g \tag{36}$$

As ϵ_h is proportional to the cohesion energy W_c, it may be shown that

$$T_g = [\Delta H' + k_1 W_c]/(-R \ln f_g) \tag{37}$$

where $\Delta H' = \Delta H_f - \Delta H_n$, ΔH_f is the heat of fusion, and ΔH_n is that part of the heat of fusion associated with hole melting. The connection between fractional free-volume and energy of hole formation, or microvoid formation, was discussed by Bartenev and Sanditov[35]. On the basis of a quasi-crystalline model of liquid it was found that

$$f = \exp \left(-\frac{\epsilon_h + \nu_h P}{kT} \right) \tag{38}$$

where $\epsilon_h = P_i \nu_h$, P_i being inner pressure and P pressure. It follows from this equation that

$$\frac{\epsilon_h + \nu_h P}{kT_g} = \ln (1/f_g) \tag{39}$$

A very interesting development of the free-volume concept has been made in some papers by Miller[29, 36]. His main idea is that all excess thermodynamic functions

$$\Delta X = X_{\text{liquid}} - X_{\text{crystal}} \tag{40}$$

may be extrapolated to zero at temperatures above 0 K[36]. At these temperatures excess volume equals zero. Miller has shown both excess entropy and excess volume become zero at the same temperature, corresponding to temperature T_0 in Vogel's equation

$$\lg \eta = \lg A + B (T - T_0) \tag{41}$$

Coefficient A is a function of the weight-average number of bonds in macromolecule Z_w:

$$\lg A = \lg K + 3.4 \lg Z_w \tag{42}$$

From Eqs. (41) and (42) it follows that at T_g

$$\lg \eta_g = \lg K + 3.4 \lg Z_w = B/(T_g - T_0) \tag{43}$$

Using the corresponding values for vinyl polymers, the following relationship can be obtained:

$$B/(T_g - T_0) \cong 15 \qquad (44)$$

From the point of view of free-volume

$$\lg \eta \propto b/\phi \qquad (45)$$

where $\phi = \Delta\alpha\,(T - T_0)$ expresses the free-volume. It follows from this that

$$\phi_g/b = (T_g - T_0)/2.3\,B = 0.029 \qquad (46)$$

which is in agreement with the WLF value. Thus the concept of an iso-free-volume state follows directly from the rheologic results. The distinction between Miller's definition and that of Simha and Boyer is that

$$\phi_g = \Delta\alpha\,(T_g - T_0) = 0.11 - \Delta\alpha\,T_0 \qquad (47)$$

where the last term is not constant for different vinyl polymers. Therefore

$$0.11/\phi_g = T_g/(T_g - T_0) \qquad (48)$$

Miller[37] supposed that the right side of Eq. (48) was equal to the number of mono-meric units in a so-called "cooperative unit" Z_g^* at T_g. In this case

$$Z_g^* \, \phi_g = 0.11 \qquad (49)$$

This condition may be regarded as an additional criterion of the glass-liquid transi-tion. According to this definition, the greater the free-volume for segmental motion the smaller the cooperative unit necessary for such motion and for the maintenance of isomobility at T_g.

A similar concept was proposed by Bueche[38], who discussed some fluctuations in free-volume and maximum free-volume f_c to maintain the motion of a group con-sisting of "n" units.

On the basis of the points discussed above, Miller suggested[39] that there is a relationship between Z_g^* and configurational entropy S_c of the form

$$Z_g^* \, S_c = K \qquad (50)$$

This was the first suggestion that the concept of free-volume somehow reflects changes in configurational entropy, which may be estimated by the parameter Z_g^*.

It is now generally accepted that the viscous flow of polymeric liquids is con-nected with chain segment rotation, i.e. with configurational entropy. From this point of view Miller concluded that the Simha-Boyer equation was not correct since the relative free-volume in SB theory equals zero at 0 K, not at $T = T_0$. If the latter condition is correct, K [Eq. (50)] is not a constant. Thus we see again that whether the fractional free-volume at T_g is a constant or not depends on the manner of its definition. On the basis of some data relating to the viscosity of melts at various temperatures Miller[40] further concluded that the liquid mobility of polymers depends much more on thermal factors than on volume, and hence that any variations of the free-volume theory combining viscosity and the relative fraction of free-volume cannot be applied to real systems.

In all the cases discussed above it was believed that the free-volume played the main role in determining the properties of liquids. However, Miller[41] noted that all theories of liquids treat the intermolecular potential energy as one of the most important factors determining molecular fraction whereas free-volume reflects only the intermolecular distances, so that all the equations, including f, lack a term for energy. According to Miller[41], the kinetic energy for a hypothetical liquid at 0 K is zero, and volume v_0 is determined only by the potential energy of interaction u_0. The value u_0 must be intimately connected with the evaporation energy ϵ_0 at 0 K, therefore parameters ϵ_0 and v_0 and "zero" cohesion energy density $(\epsilon/v)_0$ provide a basis for a definition of free-volume as $v_f = v - v_0$.

It was shown that there is an empirical equation which connects the compressibility coefficient β with the free-volume fraction:

$$-\beta \cong k f^2 (1 - f) \tag{51}$$

Here $\beta = (1/v)(dv/dP)_T$ at $P = 1$ atm. It was supposed that

$$-f^2 (1 - f) \cong 5 (\epsilon/v)_0 \tag{52}$$

or

$$-f^2 = 5 \beta \epsilon_0 / v \tag{53}$$

From this point of view, both energy and volume factors are of importance for the properties of liquids and are inserted into the equations describing liquid compressibility.

Some very interesting ideas concerning the relationship between free-volume formation and the energy of one mole of hole formation were developed in detail by Kanig[42]. Kanig introduced some improvements to the definition of free-volume. On the basis of Frenkel's ideas[43] he divided the free-volume into two parts, one of which is determined only by the thermal vibrations of atoms in the lattice of a real crystal while the other is connected with inherent free-volume, *i.e.* voids and holes. It is the latter that makes possible the exchange of particles, *i.e.* the very existence of the liquid state. He introduced some new definitions of fractions of free-volume:

$$\varphi_{f_1} = \frac{N_1 V_{f1}}{N_1 V_{f1} + N_2 V_{f2}} \qquad \varphi_{f_2} = \frac{N_2 V_{f2}}{N_1 V_{f1} + N_2 V_{f2}} \tag{54}$$

where N_1 and N_2 are the numbers of holes and molecules in the system, V_{f2} is the expansion volume determined by thermal vibrations, and V_{f1} is the hole volume.

Equation (54) represents the ratio of real volume of holes and expansion volume to the total free-volume, as distinct from its usual definition as the ratio of free-volume to total volume, the latter being determined by

$$\varphi_1^* = \frac{V_1^*}{V_1^* + V_2^*}$$ (55)

where V_1^* is the partial volume of holes and V_2^* the partial volume of molecules at T_g, including expansion volume. For this case it may be shown that

$$N_2 V_{f2} = T_g(dV_g/dT)$$ (56)

and

$$N_1 V_{f1} = T_g(dV_1/dT) - T_g(dV_g/dT)$$ (57)

where V_g is the system volume below T_g and V_1- is the same above T_g. The free-volume is determined as follows:

$$N_1 V_{f1} + N_2 V_{f2} = T_g(dV_1/dT)$$ (58)

On the basis of many experimental data Kanig concluded that the fractional free-volume at T_g as usually determined is not a constant, and therefore it is desirable to introduce additional characteristics of the iso-free-volume state. Such characteristics may be represented by the values given by Eq. (54) relating to glass temperature, φ_{f1}^* and φ_{f2}^*. In this case it follows that

$$\varphi_{f1}^* = \frac{v_{0g} - v_{01}}{v_g - v_{01}} \, 0.64 \quad \text{and}$$ (59a)

$$\varphi_{f2}^* = \frac{v_g - v_{0g}}{v_g - v_{01}} \, 0.36$$ (59b)

where v_{0g} and v_{01} are the volumes of glass and liquid extrapolated to 0 K.
    Indeed, these values appear to be more constant for various polymers than those commonly calculated. Fractions φ_{f1}^* and φ_{f2}^* are, according to Kanig constants that do not depend on the nature of the polymer and may characterize the corresponding state at glass temperature. From Eqs. (59) it follows that at T_g the ratio of hole volume to expansion volume must be constant, the latter being determined by trans-lational movements of molecules. In particular, these values are present in the equation describing T_g as a function of degree of polymerization.
    Later Kanig[44] took into account the temperature dependence of the fractional free-volume, calculated according to his equations. Below T_g, as temperature de-creases, φ_{f1}^* increases as a result of a decrease in expansion volume at "frozen" hole volume. Above T_g, φ_{f1} increases due to the sharp rise in hole volume. At T_g the value φ_{f1}^* is at its minimum. From the condition $d\,\varphi/dT = 0$ it may be found that

$\varphi_{f1}^* = (\alpha_1 - \alpha_g)/\alpha_1$. This yields practically the same value for φ_{f1}^* as is found without taking into account the temperature dependence. However, Kanig believed that the extreme values φ_{f1}^* and φ_{f2}^* correspond to the equivalent states at T_g, because these values are constants and do not depend on the nature of the substance.

Kanig's discussion[45] of the relations discussed above indicated that the traditional definition of fractional free-volume according to WLF theory may agree better with experimental results if a geometric parameter a is introduced. Kanig proceeds from the standpoint that the fractional free-volume at T_g, $\varphi_1^* = f_g$, may be considered a constant only as a very rough approximation. For many real systems it was found that polymers with more flexible chains have smaller φ_1^* values than rigid-chain polymers. The introduction of parameter a is necessary to characterize the corresponding states. The thermodynamic consideration used to justify the introduction of a is that of assuming a polymer melt to be a mixture of chain molecules 2 with holes 1 on the assumption that the melt is saturated with holes at any temperature. Taking into consideration the energy of twin interactions w_{11} (hole-hole), w_{12} (hole-polymer) and w_{22} (polymer-polymer), it becomes possible to calculate the partial thermodynamic values of the dilution of the melt with holes. These expressions contain as a parameter the values of the contact areas O_1 of hole mole and O_2 of monomeric unit mole. It can be shown that for mixing n_2 monomeric units with n_1 mole of holes,

$$\frac{1}{\varphi_{f1}} \frac{n_2 O_1}{n_1 O_1} \approx \frac{(O_2/V_2)n_2 V_2}{(O_1/V_1)n_1 V_1} \approx \frac{1}{a}(1/\varphi_1) \tag{60}$$

where V_1 and V_2 are molar volumes of monomeric units and holes and

$$a = (O_1/V_1)/(O_2/V_2) \tag{61}$$

These relationships are valid if we take $n_1 V_1 \ll n_2 V_2$ and $\varphi_1 = n_1 V_1/(n_1 V_1 + n_2 V_2)$, $\varphi_2 = 1 - \varphi_1$. In this case a has a physical meaning as a parameter that determines the size and shape of the macromolecule. This parameter, according to Kanig, is more realistic for describing macromolecules that the usually accepted coordination number. On the basis of these initial ideas, the author gives the relationship for glass temperature:

$$T_g = \frac{(1-a\varphi_1^*)H_{0i}^*}{R(\ln \varphi_1^* + \varphi_2^*)} = \frac{\Delta H_1^*}{R(\ln \varphi_1^* + \varphi_2^*)} \tag{62}$$

or $T_g = K(\varphi_1^*) \Delta H_1^*$ \hfill (62 b)

where $H_{0i}^* = O_1 \Delta w_{12}$, $\Delta w_{12} = -(1/2)w_{22} + w_{12}$, and * relates to T_g. In this manner the glass temperature becomes associated with free-volume and energy of hole formation, and therefore depends on the intermolecular energy w_{12} and the geometric factor a. Using the procedure developed, Kanig has shown that

$$T_g \, \Delta\alpha = K(\varphi_1^*, a) = \cfrac{\varphi_1^*}{2a\varphi_1^* + \cfrac{(\varphi_2^*)^2}{\ln \varphi_1^* + \varphi_2^*}} \tag{63}$$

It is understood that this equation should correspond to the empirical SB equation if K were a constant, i.e. if φ_1^*, in accordance with iso-free-volume theory were a constant value and parameter a the same for all polymers.

The advantage of Kanig's argument over other concepts consists in the establishment of a correlation between values φ_1^* and a and certain parameters that are easily found experimentally: $\Delta\alpha$, C_p (change in the heat capacity) at glass temperature, and molar cohesion energy at T_g. This allows both values, φ_1^* and a, to be calculated from the experimental data. The results, taking into account the iso-free-volume concept, give the average values

$$f_g = \varphi_1^* = 0.0235 \pm 0.0050 \text{ (average error } \pm 21\%) \text{ and}$$

$$a = 3.15 \pm 0.35 \text{ (average error } \pm 11\%)$$

It is apparent that only the average value corresponds to the universal value in WLF theory, and that the deviations from this value can be very considerable. In spite of this, we may believe that free-volume is determined mainly by the hole volume. At the same time it follows from Eq. (63) that the SB constant is a very complex value and a function of φ_1^* and a. Plotting the experimental data in coordinates

$$\Delta\alpha = \cfrac{\varphi_1^*}{2a\varphi_1^* + \cfrac{(\varphi_2^*)^2}{\ln \varphi_1^* + \varphi_2^*}} \quad \cfrac{1}{T_g} \tag{64}$$

shows an essential deviation from linearity. The variability of φ_1^* forces us to conclude that the concept of an iso-free-volume state does not pass a strict test.

4. The Iso-Free-Volume State at the Glass-Transition Point

The idea that the fractional free-volume at glass temperature as found experimentally depends on the mode of molecular motions was put forward in 1967[46, 47] as a result of calculating f_g from data obtained from isothermal volume relaxation for some polymer systems. By estimating average relaxation time at different temperatures it was possible to find the fractional free-volume f_g at T_g according to WLF theory. If we accept the validity of the theory as regards the universal dependence of the reduction factor a_T on $(T - T_g)$, then on the basis of data on $\Delta\alpha$ and theoretical values a_T calculated from universal values of the coefficients C_1^g and C_2^g, it is possible to make an estimate of f_g. In this case the value found corresponds to the universal one. If, however, we use the experimental values a_T, the fractional free-volume, calculated according to WLF theory, being constant, is much higher than the universal value, viz. 0.08–0.09. To explain this finding, it is necessary to bear in mind that in WLF theory f_g is calculated from the data for dynamic mechanical properties, when all relaxation processes are connected with segmental mobility. In this case f_g for many polymers is really quite close to the universal value and may be assumed to correspond to the hole size necessary for jumping small structural units, so determining the relaxation properties in the given frequency range. However, it had been found[48] that there are some isophase transitions in polymers associated with the mobility of some structural elements larger than segments, including a certain kind of mobility of elements of the supermolecular structure. It is evident that the motion of large structural elements needs larger holes or greater free-volume. Therefore it is necessary to accept that, having investigated various relaxation mechanisms, it is possible to determine the free-volume fraction necessary for jumping structural elements of different sizes. This follows from the data for volume relaxation corresponding to processes with large relaxation times. It is evident also that at the same molecular packing density (volume of intermolecular voids) the real free-volume for realization of a certain kind of molecular motions will be determined by the mode of these motions. When the free-volume is sufficient for mobility of small structural units, for example, segments, but insufficient for the motion of larger units, the well-known discrepancy between T_g found by different methods is observed. The same idea was suggested by Sanchez[49]. He found theoretically that average hole size increases with increasing T_g. He believed this phenomenon to be connected with the fact that polymers with higher T_g have more rigid chains, their packing density less dense than for flexible chains, so that the average hole size increases. For example, for low-molecular-weight compounds the fraction of free-volume at T_g is less than with polymers because the greater molecular mobility of small molecules needs a smaller free-volume.

According to [50] the SB rule cannot be correct, as the $\Delta\alpha$ values for different systems, where T_g may be essentially different, are very close (2.5–$4.0 \cdot 10^{-4}\,deg^{-1}$). However, Simha and Boyer[51] believed this conclusion to be mistaken due to incorrect treatment of the experimental data.

The picture given within the framework of SB theory was so unsatisfactory that new ways were sought to improve it. Simha and Weil[52] attempted this by taking into account the temperature dependence of the coefficients α_g and α_l, using extrapolation for liquids at temperatures below T_g and the corresponding functions of volume and temperature. The systematic decrease of the value $\Delta\alpha T_g$ in polymers with increasing length of side chain served as justification for this consideration. The decrease of $\Delta\alpha T_g$ must lead to a considerable decrease in free-volume with decreasing temperature. According to[52], it was necessary to revise the postulate of constant free-volume at T_g. Taking into account the temperature dependence of α_g and α_l, Simha and Weil gave alternative expressions for the free-volume fraction f in terms of volumes and expansivities, respectively:

$$f_g = 1 - \exp - \left[\int_0^{T_g} (\alpha_1 - \alpha_g)\,dT \right] \equiv (v_{0g} - v_{01})/v_{0g} \tag{65}$$

where v_{01} and v_{0g} are the extrapolated volumes of liquids and glass according to the curvature of all temperature variations. It follows that f_g is independent of temperature for $0 \leqslant T \leqslant T_g$, since the second equality remains valid throughout this range. Thus, f_g may be considered as a packing characteristic of the glass at absolute zero. The first equality may be formally extended to $T > T_g$ by extending the upper limit of integration, which requires an extrapolation of the glassy state into the liquid region.

The alternative occupied volume is simple v_{01} and, hence

$$f_g = 1 - \exp \left[-\int_0^{T_g} \alpha_1 \, dT \right] \equiv (v - v_{01})/v \qquad (66)$$

Equations (65) and (66) of Simha and Weil provide the most general expression of the SB rule and free-volume ratios as a function of the thermal expansivities α_1 and α_g. The authors have shown that, since $\alpha \leqslant 8 \cdot 10^{-4} \deg^{-1}$ and $T_g < 5 \cdot 10^2$ K, for most polymers the exponentials may be expanded to yield:

$$f_g \approx \int_0^{T_g} (\alpha_1 - \alpha_g) \, dT \left[1 - (1/2) \int_0^{T_g} (\alpha_1 - \alpha_g) \, dT \right] \qquad (67a)$$

and

$$f_g \approx \int_0^{T_g} \alpha_1 \, dT \left[1 - (1/2) \int_0^{T_g} \alpha_1 \, dT \right] \qquad (67b)$$

From Eq. (67), disregarding quadratic terms, we may obtain

$$(\alpha_1 - \alpha_g) \, T_g \cong f_g + [\alpha_1 \, n/(n + 1) - \alpha_g m \, (m + 1)] \, T_g \qquad (68)$$

where n and m are constants and the values α_1 and α_g are taken at T_g. Thus, the fact that the left-hand side is constant does not imply the same with respect to f_g, and vice versa.

As Eqs. (66) and (67) involve integration of the function $\alpha_1(T)$ over an experimentally nonrealizable region $0 \leqslant T \leqslant T_g$, the authors have given a method for the required extrapolation using reduced temperature-volume functions.

When Eqs. (65) to (67) were applied to the experimental data, a systematic decrease in the product $(\alpha_1 - \alpha_g) \, T_g$ was found to occur with increasing length of side chain in the methacrylate series. It was shown also that the value $(v_{0g} - v_{01})/v_{0g}$ is not a constant at T_g whereas the ratio $(v - v_{01})/v$ at T_g is comparatively constant. The principal discrepancies between individual systems at T_g are associated with the contribution of the glassy state to the free-volume value. The variation of f_g for different polymers may be reduced by taking into account the nonlinear dependence of the specific volume of polymers in the liquid and glassy states. The decrease of f with decreasing temperature follows also from the data[53]. The entropy of the mixing of holes with segments was considered, taking into account the temperature dependence of hole concentration to minimize the free energy. In the Gibbs-DiMarzio theory[54], the temperature T_2, at which the system is in equilibrium, and the product $\Delta\alpha T_2$ are functions of the free-volume, v_f, alone. Its constancy at $T = T_2$ corresponds to the condition $\Delta\alpha T_g$ = const, if T_2 is proportional to T_g.

In the light of the above results the concept of iso-free-volume at T_g cannot be considered proved, and indeed, it is in contradiction with many experimental data. This is why some authors[41, 44] have developed new ideas according to which liquids and polymers at T_g are in a state of isoviscosity (isoviscous state), not necessarily in the free-volume state. In connection with this we should like to note that some improvements to the SB rule were proposed[55] in the form:

$$(1/2) [(dv/dT)_l + (dv/dT)_g] T_g/v_g = K_3 \tag{69}$$

The expressions in brackets are the expansivities above and below T_g. The constant K_3 is a function of bond type in chains and is really constant for every class of polymers. The physical interpretation of this equation may be consistent with the iso-free-volume concept. However, we believe that the introduction of this equality is in practise a denial of the concept. There are also other arguments against this concept. Kästner[56] found, for example, that dielectric losses diminish during the isothermal volume contraction, which indicates a dependence of relaxation times on free-volume. However, if we assume that relaxation time depends exclusively on free-volume, the calculated reduction factor differs from the experimental one.

Some serious objections to the iso-free-volume concept were also given in[57] on the basis of data[58]. It was shown that β relaxation in polymethylmethacrylate and polyoxyprolyleneglycol proceeds in the same way at constant pressure and constant volume. It is very difficult to correlate these results with a concept according to which changing viscosity and average relaxation time are determined mainly by changes in volume. To uphold this concept it is necessary to suggest the existence of a great negative expansivity of occupied volume.

Some deviations from the Simha-Boyer rule with increasing side chain length were discovered by Simha[59]. In [50] an attempt was made on the basis of a good deal of literature data to establish the range of applicability of the SB equation. The dependence of ΔT_g and $\Delta\beta T_g$ on T_g was constructed, where

$$\Delta\alpha = (1/v) [(dv/dT)_l - (dv/dT)_g] \text{ and}$$

$$\Delta\beta = [(dv/dT)_l - (dV/dT)_g]$$

It was found that $\Delta\beta T_g$ and $\Delta\alpha T_g$ are not constant and therefore the SB equation has limited applicability. The results indicate an increase in $\Delta\alpha T_g$ with increasing T_g. Therefore it is inadmissible to use the product $\Delta\alpha T_g$ as a universal value in any theoretical discussion of the glass-transition phenomenon. At the same time, this conclusion in no way excludes the free-volume theory and the role of free-volume in the transition from the glassy to the liquid or rubberlike state.

To prove the concept of iso-free-volume, some attempts were made to determine f_g directly from experiment. In [60] free-volume was determined from the difference in specific volume of a polymer in bulk and in solution. It was found

that for polymethylmethacrylate $f_g = 0.0261$. In another paper[61] f_g determined
from the temperature dependence of specific volume was found to be 0.025 for
linear polyethylene whereas for branched PE, or PE with grafted chains of acry-
lonitryl, $f_g = 0.127$. A consideration of the equation of state of polymers[62] led to
the conclusion that agreement between experimental data and the universal value
$\Delta \alpha T_g$ is only half-quantitative. It was also shown[63] that the dependence of T_g on
pressure, which is predicted according to the free-volume concept, is not observed
for many polymers. Calculations have shown that the occupied volume is much
larger than the crystalline volume. The excess volume of glassy polymer is believed
to be connected with the existence of voids between the polymeric chains. These
voids are too small for the segmental motion but facilitate some local movements in
the glassy state, which leads to the appearance of secondary transitions for polymers below T_g.

In [60] the fractional free-volume was calculated according to the relationship:

$$T_\infty = T_g - \frac{f_g}{\alpha_1 - \alpha_g} \tag{70}$$

where T_∞ is the temperature at which viscosity approaches ∞. It was claimed that
this equation was more correct as it does not contain the arbitrary value B of the
WLF equation. In spite of this, it was found that for PMMA $f_g = 0.0048$. In [64] as
well lower f_g values were found for N-alkanes and PE than for vinyl polymers.

A very interesting approach to the theoretical prediction of the limit value
$\Delta \alpha T_g$ was developed in [49]. As a result of a thermodynamic consideration of mixing macromolecules with holes and taking into account the free-volume concept, it
was shown that

$$T \alpha_h = - \frac{E_h/RT}{\exp(1 - 1/r + E_h/RT) - 1} \tag{71}$$

Here α_h is the fraction of thermal expansion connected with changes in hole concentration (free-volume expansion), E_h is the energy of hole formation,
$r = M/\rho_0 V_n N_A$, where N_A is the Avogadro number, M molecular weight, ρ_0 the
density of a liquid without holes at absolute zero, and V_n the hole volume. For
polymeric systems r is very small, and then $\alpha_h T$ is the function of E_h/RT alone. The
value α_h is identified with experimentally observed changes in the thermal expansion
coefficients $\Delta \alpha$ at T_g, i.e.

$$T_g \alpha_h = T_g \Delta \alpha \tag{72}$$

A plot of Eq. (71) shows that the predicted T_g passes through a maximum value of
about 0.159 at $E_h/RT = 0.841$. Nearly all the values for $T_g \Delta \alpha$ reported in the literature are below this theoretical limit. The predicted $T_g \Delta \alpha$ varies very slowly with
hole energy. For E_h/RT values between 0.5 and 2.0 we have

$$0.105 < T_g \Delta \alpha < 0.159$$

It was shown that the change in the thermal expansion coefficients $\Delta \alpha$ that occurs at T_g can be predicted from the WLF theory C_2 constant $(T_\infty \equiv T_g - C_2)$,

$$\Delta \alpha = 2 \left(T_\infty / T_g \right) \exp \left[(1 + 2 \, T/T_g) - 1 \right]^{-1} \tag{73}$$

It was postulated that $T_2 = T_\infty$, where T_2 is the second-order phase-transition temperature according to Gibbs-DiMarzio theory[54]. The free-volume is determined here as $v_f = (v - v_\infty)/v_0$, where v_∞ is the macroscopic volume of the liquid at T_∞ and v_0 is the hard core volume of molecules.

The theory predicts an iso-free-volume state at T_∞, which is equal to T_0 of the Vogel equation, but an examination of the calculated values of f_g does not indicate that T_g is an iso-free-volume state.

Other cases of deviation from the behavior predicted within the framework of the free-volume concept have been reported [65, 66]. We have already mentioned Kanig's conclusion about the dependence of f_g on chain flexibility[45] and our own conclusion[46, 47] that it depends on packing density. In connection with this it is worth discussing whether there is really any dependence of the fractional free-volume at T_g on the molecular parameters of chains.

5. Dependence of the Free-Volume Fraction at Glass Temperature on the Molecular Parameters of Linear Polymers

We have already noted that the presumed invariance of f_g for polymers with widely different molecular structures is at best only a rough approximation. The deviation from constancy of f_g among the various polymers appears to be systematic and therefore should not be attributed to experimental uncertainties in obtaining these values, but rather might be related to intrinsic differences in the packing modes of chain molecules in the bulk polymer. On the basis of qualitative arguments and a discussion of the glass transition in some polyurethanes in terms of the Hirai-Eiring theory, it was proposed[67] that experimentally found values connected with the fractional free-volume for different polymers might bear a definite relationship to their molecular parameters, e.g. chain flexibility. Now we will analyze this suggestion in detail and show that the absolute value f_g might be related to the efficiency of molecular packing of a polymer. It is convenient for this purpose to use Litt and Tobolsky's definition of occupied volume, that is v_3'' (see Fig. 1). From an examination of the data given in Table 1, where the molecular parameters of some polymers are collected, it is readily seen that there is a definite trend for the ratio v_3''/v_g to increase with increasing bulkiness of the side groups of polymer chains. We can rationalize this fact by applying the ratio between polymer volume in the crystalline and amorphous state, v_c/v_a, to the characteristic chain parameter σ [68], where a is the chain thickness obtained from the crystalline unit cell dimensions and σ is the steric factor, or flexibility parameter for an isolated chain, which is usually determined by mea-

Table 1. Representative values of T_g, V_3'', V_g, a, σ and $\Delta\alpha$

Polymer	V_3'' cc/g	V_g cc/g	T_g °K	σ	a A	$(K_c)_g$	$\Delta\alpha$ deg^{-1}
1. Polyethylene	0.967	1.055	160	4.25	1.63	0.764	
2. Polypropylene	1.050	1.120	238	6.0	1.87	0.724	4.2
3. Polyclorotrifluoroethylene	0.460	0.525	318	5.6	2.03	0.746	
4. Polystyrene	0.920	0.985	360	8.4	2.3	0.695	3.0
5. Polymethylmethacrylate	0.830	0.870	318	7.95	2.14	0.705	3.64
6. Poly-4-methylpentene-1	1.230	1.195	302	9.3		0.665	3.78
7. Polyisoprene 1,4-cis	0.975	1.03	201	5.3	1.67	0.742	
8. Polyisobutylene	1.02	1.04	203	6.45	1.8	0.726	4.7
9. Polydimethylsiloxane	–	0.852^c	150	7.5	1.47	–	9.3
10. Polyethylene oxide	0.739	0.825	206	4.35	1.63	0.818	5.95
11. Polypropylene oxide	0.885	0.945	198	4.95	1.62	0.685	6.0
12. Polytetramethyleneoxide	0.885	0.960	187	4.45	1.68	0.769	6.4
13. Polyethylene adipate	0.735	0.790	216	4.45	1.68	0.772	6.1

suring its intrinsic viscosity in an ideal solvent. The relation between v_c/v_a and a/σ was shown to exist [67]. The relevant values of the parameters a and σ are also given in Table 1. It was found that there was a definite trend for the ratio v_3''/v_g to increase with a/σ.

It can be shown that the ratio v_3''/v_g is equal to the ratio of polymer packing densities coefficients in the amorphous and crystalline states, K_a/K_c at T_g, because, by definition, $K_a = N_A v_i/v_a$ and $K_c = N_A v_i/v_c$, where v_i is the Van der Waals volume of the chain repeat unit. The calculated values of $(K_c)_g$ correlate with the characteristic chain parameter a/σ, the relationship between them being expressed by a linear equation

$$(K_c)_g = 0.9 \, (1 - 0.058 \, a/\sigma) \tag{74}$$

Summarizing, we can say that the free-volume fraction at T_g as defined by Eq. (12) diminishes as the chain packing in the bulk polymers loosens. This result confirms the analogous qualitative conclusion formulated earlier [46, 47].

We believe that the conclusion reached is of general significance for all polymers in the amorphous state. However, Eq. (74) cannot readily be applied to noncrystallizable polymers, because in this case the value of v_3'' or of a becomes obscure. In attempting to extend the main ideas put forward above to cover this case, we employ an alternative definition of f_g, that is

$$f_g = (\alpha_1 - \alpha_g)(T_g - T'') \tag{75}$$

where T'' is a reference temperature which is related to the occupied volume v_0'', v_2'' and v_3'', as shown in Fig. 1. For example, $T'' = 0$ K corresponds to SB theory and $T'' = T_3$ to Litt and Tobolsky's definition, while $T'' = T_2$ coincides with WLF theory.

Earlier[69] we obtained the following empirical equation relating T_g to the chain stiffness parameter:

$$T_g = A \, (\sigma - b) \tag{76}$$

in which A and b are constants. Equation (76) is compatible with the SB rule [Eq. (23)], if we suppose a proportionality to exist between A and K_1, and between $\Delta\alpha$ and $(\sigma - b)^{-1}$ in Eq. (76). It can be qualitatively shown [69] that this proportionality is at least conceivable. The coefficients of thermal expansion of a polymer may be related to the conformational characteristics of isolated chain as well. It seems feasible to postulate proportionality between the said parameters in Eqs. (23) and (76) and, if this situation is real, the constant K_1 in Eq. (23) will not necessarily be the same for all polymer series[69].

Solving Eqs. (23) and (76) for T_g and inserting the numerical values of the parameters involved, we obtain

$$\Delta\alpha = 4.2 \cdot 10^{-4} \, (\sigma - b)^{-1} \quad \text{and}$$

$$\Delta\alpha \cdot \sigma = \Delta\alpha = 4.2 \cdot 10^{-4} \tag{77}$$

As can be seen from Eq. (77), the quantity $\Delta\alpha$ is definitely not a constant but should grow with an increase in chain flexibility (*i.e.* decrease of σ). Thus, if both the SB relationship and our Eq. (76) are correct, the plot of $\Delta\alpha$ against $\sigma \cdot \Delta\alpha$ should be a straight line with a slope of unity. And, indeed, $\Delta\alpha$ actually tends to increase with decrease of σ, but this behavior cannot be quantitatively accounted for by Eq. (77), as the deviations of the experimental points from the theoretical line become steadily more pronounced as $\Delta\alpha$ decreases. The best fit of experimental values of $\Delta\alpha$ is observed only for those polymers which exactly conform to the SB rule (polyvinylchloride, polyvinylacetate, etc.). This result emphasizes once more that the value of f_g as defined earlier is not a constant, at least for the many varieties of polymers. To gain a deeper insight into the expected relationship between $\Delta\alpha$ and σ, we plotted lg $\Delta\alpha$ against $\sigma\Delta\alpha$. In this case much better linear dependence clearly results, with the experimental points fluctuating around the theoretical line even for those polymers which do not obey the SB rule.

It is possible to give some quantitative explanation of the relationship between $\Delta\alpha$ and the parameter σ. By definition, α_g is determined by the anharmonic vibrations of isolated molecules or certain molecular groups, which suggests that the physical nature of α_g is nearly the same for polymers and liquids. On the other hand, the nature of α_l cannot be the same. For liquids, α_l decreases when intermolecular forces increase whereas T_g increases[24]. Therefore we may expect the product $\Delta\alpha T_g$ to be constant for liquids. For polymers, however, α_l is dependent not only on the inter-molecular interactions influencing the average distance between the centers of gravity of molecular groups, but on the ability of macromolecules to make confor-mation changes as well. For this case, for polymers, we can write:

$$\Delta\alpha = (\alpha_l)_n - \alpha_g = (\alpha_l - \alpha_c)_m \cdot \Delta\alpha_{conf} \tag{78}$$

where the indices m and n denote polymer and monomer and $\Delta\alpha_{conf}$ is an additional contribution to $(\alpha_l)_n$, connected with a change in the number of conformations in the melt[70]. It is clear that the numerical value of the coefficient $\Delta\alpha$ for polymers will depend on the value of the "polymer" part, $\Delta\alpha_{conf}$, which will obviously be different for polymers with differing chain flexibility.

The conclusion that the free-volume fraction at T_g is not a universal parameter for linear polymers of differing molecular structure can be qualitatively confirmed by the following arguments[71]. Assume that at temperatures far below T_g polymeric chains are in a state of minimum energy of intramolecular interaction, *i.e.* the frac-tion of higher-energy ("flexed") bonds is zero[54]. On the other hand, let the equi-librium fraction of flexed bonds at $T > T_g$ obey the Boltzmann statistics and be a function of Boltzmann's factor ϵ/kT. Thus, the fraction of flexed bonds at T_g can be estimated from the familiar expression:

$$f_g = [\exp(-\epsilon/kT_g)]/[1 + \exp(-\epsilon/kT_g)] \tag{79}$$

It follows from this equation that f_g decreases as the ratio $\epsilon/kT = c$ increases. Taking into account the obvious fact that the volume increment per flexed bond should be larger than that for an unflexed one, the above result means that the free-volume fraction at T_g will generally tend to decrease with the decrease of polymer packing density in the crystalline state (here the fraction of flexed bonds f_g may be identified with fractional free-volume). The treatment of some experimental data to find the dependence of T_g on chain flexibility[71] has shown that, on the basis of the general equation (Eq. 76), it is possible to obtain an expression for linear polymers linking T_g with free-volume in the form

$$T_g = T_g(\text{LPE}) + [\sigma - \sigma(\text{LPE})] \, Y(f_g) \tag{80}$$

where LPE denotes linear polyethylene and $Y(f_g)$ is a function to be evaluated from the experimental data. These results can also be considered as showing the necessity of some revision of the famous concept of T_g as an iso-free-volume state.

It is worth noting that f_g decreases with increasing a/σ. This change is brought about mainly by variation of the polymer packing density in the crystal, $(K_c)_g$, whereas the value of the packing coefficient in the amorphous state, $(K_a)_g$, is nearly invariable. It seems reasonable, therefore, to suggest that it is the value $(K_a)_g$ which should be taken as a measure of the free-volume fraction at T_g. To put it in another way, the definition of the occupied volume as the intrinsic or Van der Waals volume of the chain repeat unit with a small (close to zero) thermal expansion coefficient appears the most appropriate. This proposal is not inconsistent with one of the possible definitions of free-volume as discussed by Bondi[73]. Moreover, our definition closely resembles the one stated by Simha as a third-law type of expansion of the occupied volume solely by the mechanism of thermal vibrations[52]. The relationship between f_g and the packing coefficient can also be written in the following form:

$$K = \left(1 - \frac{kT_g}{E_h}\right)(1 + 1/\ln f_g)\,0.66 + 0.75 \tag{81}$$

where $K = Vc/V$, V_c being intrinsic volume, and V real volume[74]. It is clear that f_g does depend on the mode of packing, which in turn is connected with chain flexibility.

6. The Concept of Free-Volume Distribution

From the point of view of the ideas discussed above concerning the variability on the free-volume fraction at T_g, even for the same modes of molecular motion in different polymers, there is great interest in some new concepts about the free-volume distribution, in the system, first proposed in [24]. The starting point is the suggestion that all molecular motions, like transfer phenomena, can take place only when the size of the voids or holes in the system exceeds a critical value v^*. This critical volume appears as a result of redistribution of the free-volume within the system. It is supposed that this redistribution proceeds without energy consumption. The idea put forward was that excess volume $\bar{v} - v_0 = \Delta\bar{v}$ ($\bar{v}$ is the specific volume, v_0 is the volume of molecule) and free-volume v_f are related as follows:

$$\bar{v} = v_f + \Delta v_c \tag{82}$$

It was supposed that only part of the free-volume may be redistributed without energy consumption, and that this takes place at a temperature above the critical value T_2. In this case the thermal expansion of the amorphous phase is determined only by anharmonic vibrations of the molecules. It follows that $v_f \approx 0$ at $T < T_2$ and $\Delta \bar{v} \approx \Delta v_c$. It is assumed also that in this case the increase in entropy due to volume change is very small. With increasing temperature a value $\Delta \bar{v}_g$ is reached, after which the main contribution in expansivity gives the expansion of the "free-for-redistribution" volume. Here $v_f = \alpha \bar{v}_m (T - T_2)$, where α is the average value of expansion coefficient and $\bar{v}_m$ is the average volume of molecules in the temperature range T_2 to T_g. The free energy of the amorphous phase should be at its minimum when the free-volume is distributed in an arbitrary way. Such a distribution may be inherent only in the amorphous phase, not in the crystalline phase. When $\Delta \bar{v} > \Delta \bar{v}_g$, at equal volumes the amorphous phase is more stable than the crystalline phase because of the positive configurational entropy connected with the free-volume distribution. Therefore, according to [24] the glass temperature of the amorphous phase is defined as a temperature of excess volume equal to $\Delta \bar{v}_g$, above which the free volume emerges.

The probability that the free volume at the given temperature exceeds the value $v*$, according to [14], is

$$p(v*) = \exp(-bv*/\langle v \rangle) \tag{83}$$

where $\langle v \rangle$ is the average free-volume per molecule, i.e. the total free-volume divided by the molecule number, and b is a numerical factor close to unity. It is possible to introduce another definition, the average free-volume in a volume unit (average fractional free-volume) f. If we denote $bv* = B$, then

$$p(B) = \exp(-B/f) \tag{84}$$

These ideas were used to describe the diffusion of low-molecular-weight substances into polymers[75]. The molecular mobility m_d in polymeric media will depend on the probability that a molecule is positioned next to a hole of size sufficient for the displacement. If we denote by B_d the value B corresponding to the minimum hole size, then

$$m_d = A_d \exp(-B_d/f) \tag{85}$$

where A_d is the factor of size and shape of a solvent molecule. Litt[70] developed ideas about minimum possible diameter $b*$ of hole connected with the macromolecule required for the realization of molecular or segmental movements. According to this idea, only holes and cavities of diameter $\geqslant b*$ (summarizing f_{ef}) can influence viscosity. From this, we derive a modified Doolittle equation:

$$\ln(\eta/\eta_0) = 1/f_{ef} = \pi^{1/2}/[A(T*/T)^{3/2}(1 - T*/T)^{-1} \exp(T*/T)] \tag{86}$$

where $A = (4/3) \pi b^* n$, n being the number of molecules per cm^3, and $T^* = 4 \pi b^{*2} \gamma / k$ (γ is the surface tension and k the Boltzmann constant, T^* being the characteristic temperature). The value of A changes from 0.75 for flexible chains to 1.4 for rigid ones. It was found that for many systems $T_g/T^* = 0.360 \pm 0.1$. It is believed that this modification of the theory of free-volume improves its physical significance. So if, according to the WLF equation, $f \to 0$ at $(T_g - 50)$ and $\ln \eta \to \alpha$, in the case discussed f_{ef} decreases monotonously at $T < T_g$, reaching zero only at 0 K.

The necessity of dividing free-volume into its constituent parts was also shown in [76]. It was noted that the fractional free-volume according to WLF is lower than in other estimations, e.g. according to compressibility. According to the authors' data for some polymers, the free-volume fraction at T_g is 0.1–0.15 whether it is calculated from compressibility or from additivity of volume of different atomic groups. The difference between this and the WLF value is attributed to the fact that some holes lose their mobility and are in a "frozen" state, and these "frozen" holes form a "weak spot" in the structure. What the authors have in mind is not a diminishing rate of molecular rearrangement, but the total exclusion of some holes from the overall process. There is thus a need to distinguish between the geometric free-volume (about 0.1–0.15), corresponding to the SB definition, and the physical free-volume, which determines the relaxation processes (about 0.03). This free-volume v_f is determined as $v_f = v_h n^*$, where v_h is the volume of excited holes and n their number. The fraction of excited holes at glass temperature will be $n^*/n = f^*/f_{ef} = 0.1$, where n is the total number of holes in the system, f^* is the fraction of physical free-volume and f_{ef} the fraction of geometric free-volume. The glass transition from this point of view is like "destroying by frost" some characteristic frequencies and is accompanied by a decrease in free-volume and mobility. It is supposed that the molecular rearrangements are realized due to hole movements corresponding to definite kinetic units with a characteristic frequency of thermal vibration. This view corresponds to our idea that every type of molecular motion needs its own free-volume.

In reality, the data on isothermal contraction for many polymers[6] treated according to the free-volume theory show that quantitatively the kinetics of the process does not correspond to the simplified model of a polymer with one average relaxation time. It is therefore necessary to consider the relaxation spectra and relaxation time distribution. Kästner[72] made an attempt to link this distribution with the distribution of free-volume. Covacs[6] concluded in this connection that, when considering the macroscopic properties of polymers (complex moduli, volume, etc.), the free-volume concept has to be coordinated with changes in molecular mobility and the different types of molecular motion. These processes include the broad distribution of the retardation times, which may be associated with the local distribution of the holes.

Mason[77] developed ideas about the distribution of the free-volume to explain the existence of the broad transition region from glassy to rubberlike state. He believed that there is some localization of that part of the free-volume that distinguishes the rubberlike state from the true liquid state in which the free-volume is not localized. In the non-crosslinked state of some rubbers there may be an arbitrary distribution of the free-volume v_f connected with the free-volume of each monomeric unit v_f'. Taking the universal value f_g, Mason gives the expression for an average free-volume per monomeric unit as

$$\bar{v}' = Mv/N_A n \tag{87}$$

where n is the number of monomeric units in the macromolecule and v is the specific volume. The monomeric free-volume f' may be defined as

$$f' = v_f'/\bar{v}' = n N_A \, v_f'/M v \tag{88}$$

The packing geometry may be characterized by the distribution functions $p\,(f)$ representing the density of monomeric units with partial free-volume f'.

The effective T_g for a monomeric unit is determined by the condition $f' = f_g$. Thus, 1 g of rubber at temperature T contains n_g monomeric units in the glassy state:

$$n_g = \int_0^{f_g} p \, df' \tag{89}$$

Similarly, in the liquid state

$$n_1 = \int_{f_g}^{\infty} p \, df' \tag{90}$$

at the condition of normalization

$$\int_0^{\infty} p \, df' = N_A \, n/M \tag{91}$$

Hence, the expansion coefficient may be expressed as

$$\alpha = \left\{ \int_0^{f_g} p \, df' \, \alpha_g + \int_{f_g}^{\infty} p \, df' \, \alpha_1 \right\} \Big/ \int_0^{\infty} p \, df' \tag{92}$$

and

$$\alpha - \alpha_g = \frac{M\,(\alpha_1 - \alpha_g)}{n\,N_A} \int_{f_g}^{\infty} p \, df' \tag{93}$$

In Eq. (93) the coefficient α is supposed to be the sum of the expansivities of each region, taking into account their number fraction [Eq. (91)]. In deriving Eq. (93) it was assumed that at temperature T some monomeric units are in the glassy and some in the liquid state in accordance with the distribution function. The integral value, representing the number of monomeric units in the liquid state diminishes from $N_A n/M$ at the temperature above T_g where $\alpha = \alpha_1$ to zero below T_g ($\alpha = \alpha_g$). If all monomeric units have the same partial free-volume f', the transition to the liquid state will take place at a strictly defined temperature. The broadening of the transition is related to the distribution of f'. The temperature interval where α diminishes from α_l to α_g gives the measure of the width of the distribution. Thus, supposing the normal distribution f' is above T_g, the average f' will be equal to the macroscopic value f and the distribution will have the following form:

$$P = p\,(f) = \frac{1}{\sigma\sqrt{2}}\,\exp\{-(f'-f)^2/2\,\sigma^2\} \tag{94}$$

where σ is the dispersion.

Thus, taking into account Eq. (91), it follows that

$$\alpha - \alpha_g = \frac{M\,(\alpha_l - \alpha_g)}{N_A\,n\,\sigma\sqrt{2}}\,\int_{f_g}^{\infty}\exp\left[-(f'-f)^2/2\,\sigma^2\right]df' \tag{95}$$

Near T_g, σ may be accepted as a constant and then in Eq. (95) the temperature comes into integral function only through f.

Taking into account the equation derived, the author managed to calculate the distribution of the free-volume for monomeric units $p\,(v_f')$ as a function of v_f' for vulcanizates of different crosslinking density.

The idea of free-volume distribution is an essential contribution to the understanding of the mechanism of the glass-transition phenomenon in polymers and to the development of free-volume theory. The need to allow for the free-volume distribution was noted also by Kanig[45] and Sanchez[49]. However, it took more than 10 years for the new interpretation, based on the idea about hole size distribution, to emerge. Kilian[75] has discussed the glass transition from the thermodynamic point of view as a process involving the mixing of polymer chains with holes of uniform size. The mixture is considered as a saturated solution of linear chains of length y_p, where y_p is the number of repeat units in a polymeric chain. However, the voids or holes have their own size distribution. In this connection we have to consider the quasistatic glass temperature at which the glass transition proceeds according to the cooperative mechanism. The partial molar enthalpy of mixing was calculated, accounting for the zero interaction between holes and between holes and macromolecules. It is supposed that hole formation in the media requires the same permanent excess entropy. For this case author gives the equation for T_g:

$$1/T_g = \frac{R}{A_p^h\varphi_p y_p} + \frac{A_p^s}{A_p^h} + \frac{R}{A_p^h y_{li}\,\varphi_p^2}\left\{\frac{1-\varphi_p}{\bar{y}_1}\,y_{li} - \ln\varphi_{li} - 1\right\} \tag{96}$$

where $i = 1,\ldots N$, A_p^h is the specific enthalpy of any hole formation, A_p^s is the excess entropy parameter of hole formation, $\varphi_p = y_m n_m/\Sigma y_i n_i$, y_{li} is the hole size, $\bar{y}_1$ is the average particle size in the mixture, and φ_{li} is the fraction of holes of size i. From Eq. (96) it follows that for a polymeric system where there exists a hole size distribution, the volume fraction of holes y_{li} should be different and should depend on the value y_{li}. From the thermodynamic point of view, every hole fraction should have its own value y_{li}, corresponding to the condition of thermodynamic equilibrium.

Equation (96) was compared with the literature data concerning the dependence of T_g on $\bar{y}_p$. It was found that, if $y_{li} = 1 - y_p$ is taken as a normalization condition, then

$$\varphi_{li} = \left[\frac{(1 - \varphi_p)e}{(1 - \varphi_p)e + 1} \right]^{y_{li}} \frac{1}{e} \tag{97}$$

represents the distribution function for hole size. If the condition that the iso-free-volume $\varphi_1(T_{g\infty}) = \text{const}$ is fulfilled, then the distribution function [Eq. (97)] is also invariant with respect to the system.

The calculation of entropy of hole formation shows that entropy has a negative value, which means that there cannot be a statistically disordered hole distribution. However, the author believes it is problematic to introduce hole distribution according to the condition $1 \leqslant y_{li} \leqslant \infty$ without relating it to molecular structure. The distribution function for a saturated mixture is monotonic and the volume fraction of large holes diminishes with increasing y_{li}. The introduction of the idea about distribution is claimed to correspond to the statistical character of molecular conformations in the liquid phase. According to Frenkel[43], the value of the free-volume in the liquid is not related to molecular size, being comparable to atomic size. This circumstance allows us to conclude that hole distribution, at least in polymeric systems, should be nonuniform in order to realize the various types of molecular motion.

7. Free-Volume in Heterogeneous Polymer Systems

From the point of view of the applicability of the iso-free-volume concept, it is of great interest to test it for some systems more complicated than simple polymeric liquids, despite the fact that, as we have already shown, this concept has failed in many cases even for simple polymers. The more complicated cases considered are compositions and polymeric blends and alloys.

At present it is well established that the existence of the phase border between a polymer and any solid leads to the appearance of different types of micro- and macroheterogeneities at the molecular, supermolecular, and chemical levels[78]. It is established that, due to adsorption interaction at the interface, an essential decrease in molecular mobility takes place as a result of which the glass temperature of such systems increases[79]. At the same time, due to retardation of the relaxation processes in the surface layers, some loosening of packing takes place, whereas in pure adsorption layers some increase in density is observed[80].

In addition, in polymeric compositions, *i.e.* in filled polymers, as a rule a broadening of the relaxation spectra takes place[81]. All these characteristics are closely associated with molecular parameters and free-volume, therefore it is interesting to consider how structural changes, induced in the same polymer by the action of the solid surface, may influence the applicability of the free-volume concept and the approach to such systems *via* the iso-free-volume state. In [82] the applicability of the

SB rule was studied for polymethylmethacrylate filled with quartz powder, using values for α_l, α_g and T_g determined experimentally from dilatometric data. It was shown that the values $\Delta\alpha\,T_g$ are the same for all systems in spite of the increase in T_g due to filling. The constancy of the value $\alpha_l T_g$ was considerably less $(0.17–0.22)$. The approximate correspondence to the SB rule was explained by loosening of molecular packing at the interface and restriction of molecular mobility. The difference $\alpha_l - \alpha_g$ consequently diminishes, both coefficients being higher than for unfilled samples (Table 2). It is worth noting that for these tests the additivity of the coefficients of thermal expansion for both polymer and filler was accepted. The compliance with the SB rule may serve as a criterion for this additivity.

In some reports[83, 84] the change in the fractional free-volume was calculated at temperatures above T_g for epoxy resin filled with polystyrene particles on the basis of the experimental value of the reduction factor a_T and the universal value f_g according to the equation

$$\lg a_T = B \left(1/f - 1/f_g\right) \tag{98}$$

It was found that the total fraction of the free-volume in the system increases with increasing concentration of the polymeric filler. The temperature dependence of f_g for the epoxy matrix was calculated on the supposition that free-volume is an additive value of the constituent components and using the temperature dependence of the fractional free-volume of polystyrene. It was found that with increasing filler concentration the fractional free-volume becomes greater than for pure epoxy resin. Since the fraction of the free-volume increases with increasing total surface area of the filler, it may be supposed that this effect is associated with the surface layers of polymer. It was found that the rate of free-volume expansion in a filled system is higher than in an unfilled one, which means that the expansivity of the free-volume

Table 2. Expansion coefficients and universal values for some filled systems

System	T_g, K	$\alpha_g \cdot 10^{-4}$	$\alpha_l \cdot 10^{-4}$	$(\alpha_l - \alpha_g)T_g$	$\alpha_l T_g$
PMMA (low mol. wt.)	335	1.50	3.98	0.083	0.133
PMMA + 5% glass fiber	354	1.10	3.12	0.071	0.110
PMMA + 20% glass fiber	370	1.75	4.05	0.085	0.152
PMMA + 30% glass fiber	387	1.80	5.50	0.143	0.213
PMMA + 5% acrylic fiber	355	2.50	5.25	0.098	0.186
PMMA + 20% acrylic fiber	369	2.80	5.50	0.098	0.199
PMMA (high mol. wt)	372	1.50	4.75	0.121	0.177
+5% acrylic fiber	374	3.25	6.50	0.122	0.243
+20% acrylic fiber	376	3.30	6.75	0.129	0.254
Polystyrene	364	2.00	5.00	0.110	0.183
PS + 5% quartz powder	365	2.00	5.05	0.110	0.183
PS + 20% quartz powder	370	2.15	5.70	0.131	0.213
PS + 30% quartz powder	371	2.20	6.20	0.148	0.232
PS + 5% acrylic fiber	365	2.50	5.05	0.101	0.180
PS + 20% acrylic fiber	367	2.50	6.25	0.138	0.229

in the surface layer is not the same as for the pure polymer. These effects are related to the loose molecular packing in the surface layers of filled polymers. Thus the free-volume expansion appears to depend on the polymer structure, which was not previously taken into account.

Some interesting results were obtained when investigating the viscosity of the melts of filled polymers (oligoester). The temperature T_0 of Vogel's equation and the fractional free-volume $f_g = (T_g - T_0)/B$ [26] were determined experimentally[85]. It appeared that the fractional free-volume in filled systems increased in proportion to the polymer fraction in the surface layer, determined independently, and ranging from 0.025 to 0.043. This fact was explained by the diminishing molecular packing density on the surface. There was at the same time a decrease in the temperature T_0. The findings indicate that the criterion of constancy of the free-volume fraction at T_g cannot be applied to filled systems because of the influence of the filler on the polymer structure. Thus, even for one and the same polymer, the difference in its physical structure induced by physical actions capable of changing the structure causes polymer behavior to deviate from that predicted within the framework of the iso-free-volume concept.

Krauss and Gruver[86] investigated the change in expansivity in filled copolymers of butadiene and styrene (the filler being carbon black with particle sizes differing by a factor of 10). From the data obtained, they concluded that the coefficient of expansion in the presence of filler does not change above T_g, but diminishes below T_g. This effect was attributed to dilatation due to stresses set up around the filler particles and arising from the difference in the expansion coefficients of filler and polymer, the stresses not being relieved in the glassy state. Since the coefficient of thermal volume expansion is related to the free-volume of the system, the molecular mobility may change due to this factor alone causing a shift toward higher temperatures in the relaxation-time spectrum. In spite of the differing explanations of the reasons for the changes in thermal expansivity given in [82] and [86], the very fact that the coefficient changes due to physical action on the polymer is very important. It shows, as we have already noted, that this value is dependent on the character of the physical action. For the same reason, the fractional free-volume must depend on the same factors.

The dependence of f_g on filler content was carefully investigated for filled polystyrene[87] and the values for f_g were calculated in different ways, using various values for the occupied volume v_0. The results of these calculations have shown that the values for f_g do not coincide when calculated in different ways. Nor are these values constant for the different amounts of filler incorporated. This shows once more that the glass temperature is not a temperature of constant fractional free-volume.

The effects associated with the influence of the phase border are especially obvious in heterogeneous polymer systems, where both components are of a polymeric nature. Such systems include polymer blends and polymers filled with polymeric filler. These two types of systems differ in that in blends it is difficult to distinguish between the two polymers as a disperse phase and dispersion media due to uniform distribution of both components in the volume.

Let us now consider the applicability of the free-volume concept to the glass transition in heterogeneous polymeric systems consisting of two polymers. We think[88] it necessary to introduce here the concept of the localized distribution of a free-volume in the system and of the partial free-volume of various chain fragments which participate in different kinds of molecular movement. The necessity for such an approach, which differs from that utilizing the statistical distribution in [77], is believed to be especially apparent in the analysis of glass transition in microheterogeneous polymeric systems. The latter include block and graft copolymers, linear condensation polymers, and others. These systems may undergo microsegregation into distinct phases, which leads to the appearance of microheterogeneity. In this case partial values of free-volume may be attributed to each phase. As an example, let us consider the micro-heterogeneous system formed by a block polymer. In a block polymer composed of two noncrystallizable components, two distinct glass transition temperatures exist, which suggests a localized free-volume distribution in the system. In other words, the increase of free-volume in one region does not influence mobility in another phase (region). If the chain is composed of i block and consists of j rigid and $\bar{j}$ soft blocks ($i = j + \bar{j}$) then, assuming that each block has its own partial free-volume, v_j and $v_{\bar{j}}$ per g of polymer, the overall free-volume is defined as

$$v_\mathrm{f} = N_\mathrm{L} \left(v_j j + v_{\bar{j}} \, \bar{j} \right) \tag{99}$$

where N_L is the Loschmidt number. Furthermore, taking into consideration the difference in the thermal expansion coefficient of both types of region, the contribution of the partial free-volume to the overall free-volume of a system will be a function of temperature. Therefore, when we deal with the viscoelastic properties of a microheterogeneous polymeric system (for example, viscosity, which is a function of the overall free-volume of a system), the value of the free-volume fraction at either of the T_g will not correspond to the universal value. In fact, if for a two-phase system the overall free-volume is $v_\mathrm{f} = v_1 + v_2$, then at the first glass-transition temperature, $T_{\mathrm{g}1}$, the fractional free-volume is $f_{\mathrm{g}1} = v_1/v$, and at the second, $T_{\mathrm{g}2}$, $f_{\mathrm{g}2} = (v_1 + v_2)/v$, v being the overall volume of the system. (This equation does not take into account the expansion of the first free-volume in the temperature range between $T_{\mathrm{g}1}$ and $T_{\mathrm{g}2}$.) Therefore, the free-volume fractions at corresponding T_g will not be equal either to each other or to any constant value. Alternatively speaking, no universal value for a free-volume fraction at glass temperature can be used to describe the viscoelastic properties of such a microheterogeneous system.

Let us now discuss the applicability of the Simha-Boyer rule to the systems under consideration. If the system has two glass transition points, $T_{\mathrm{g}1}$ and $T_{\mathrm{g}2}$, for soft and rigid blocks, respectively, we can obtain by experiment three expansion coefficients: α_1 to $T_{\mathrm{g}1}$, α_2 between $T_{\mathrm{g}1}$ and $T_{\mathrm{g}2}$, and α_3 above $T_{\mathrm{g}2}$. After $T_{\mathrm{g}1}$ is reached, the overall expansion of the system is the sum of the free-volume expansion in regions composed of soft blocks and of the common thermal expansion of glassy regions formed by rigid blocks, and after $T_{\mathrm{g}2}$ the expansion amounts to the sum of free-volume expansion of both components. For such a case it was calculated that, if the volume fractions of the components are w_1 and w_2, and if we assume the additivity of the expansion coefficients and neglect small changes with temperature in the volume fractions of the components, then the SB rule may be written as follows:

$$(\alpha_{11} - \alpha_{g1}) T_{g1} = \Delta\alpha_1/w_1 T_{g1} = (\alpha_{12} - \alpha_{g2}) T_{g2} = \Delta\alpha_2/w_2 T_{g2} = \text{const} \qquad (100)$$

Here $\Delta\alpha_1 = \alpha_2 - \alpha_1$, $\Delta\alpha_2 = \alpha_3 - \alpha_2$, α_{11}, α_{g1}, α_{12}, and α_{g2} are the thermal expansion coefficients of pure components 1 and 2 below and above their glass temperatures. The expressions in Eq. (100) involve experimentally measurable quantities only and differ from the common SB equation in that the former include the values of the overall expansion coefficients and the volume fraction of the components. It should be emphasized that these equations apply only when a system segregates into two phases, the usual situation in the majority of block and graft polymers and polymer blends[89−91]. On the other hand, if the rigid blocks are randomly distributed in the system and serve only as a temporary crosslinks capable of dissociation at the glass temperature, as in homogeneous systems, then the SB rule may be used unchanged. It is evident that Eq. (100) in the form proposed may be used as a criterion for the existence of microsegregation in a system.

However, in many real systems both in block copolymers and in polymer blends[91] the components may mutually influence each other due to interphase interaction[90, 92]. Such interaction may cause the system behavior to derivate from that predicted within the framework of the free-volume theory for a two-phase system.

This question was investigated in [93] where the applicability of the modified SB equation was examined for mixtures of two thermodynamically incompatible polymers. Some results have been given in Table 2. As may be seen from Table 2, for PMMA filled with polyacrylonitryl there is considerable deviation of the value $\Delta\alpha T_g$ from the universal value. In [93], for all the blends investigated, the glass temperatures of the components remain unchanged and are the same as for pure components, which shows the coexistence of the two independent phases. Some data are given in Table 3. These data were obtained from the dependence of the specific volume on temperature. It is seen that the product $\Delta\alpha T_g$ for each component is close to the SB constant (excluding polycarbonate). Column 9 of Table 3 gives the values calculated according to Eq. (100). It is apparent that these values deviate considerably from the SB constant found experimentally for the same components (1.5−2 times greater). At the same time the product $\Delta\alpha T_g = K_1$ for blends is lower than the corresponding SB value. Thus it may be concluded that the system under investigation cannot be described by Eq. (100). This indicates the existence of some kind of interphase interactions, which may account for the deviations from theoretically predicted behavior. In this case, the additivity of the expansion coefficients should not be assumed. An estimation of the additivity was made using the experimental data and a positive deviation was found in all cases[93]. In [94] it was also shown for different systems that there were some deviations of expansion coefficients from additivity, and these were attributed to the intermolecular interactions that determine the compatibility of the components. The higher expansion coefficient of the components in blends points to the conclusion that intermolecular interaction in the system is less pronounced, i.e. the packing density is less than in the pure component. At the same time the molecular mobility increases. These results were explained on the basis of investigations of the isothermal contraction of blends[93]. It was found that in blends at temperatures above the T_g of one of the

Table 3. Some thermal characteristics of polymer blends

System	T_g	ΔT_g	v_g cm³/g	$\alpha_g\,10^{-4}$ grad⁻¹	$\alpha_l\,10^4$ grad⁻¹	$\Delta\alpha$	$\Delta\alpha T_g$ (K_1)	$\Delta\alpha T_g/w$ (K_2)	mol. wt 10^4
1. Polybutylmethacrylate	292	–	0,9475	3,54	6,73	3,39	0,100	–	18,4
2. Polymethylmethacrylate	378	–	0,8625	4,55	7,39	3,17	0,112	–	626
3. Polystyrene	357	–	0,9775	3,41	0,05	2,64	0,095	–	16,4
4. Polycarbonate	411	–	0,8775	4,42	9,12	4,70	0,193	–	4,5
5. PBMA + PMMA (1:1)	291	–	0,9050	4,43	6,98	2,55	0,074	0,148	–
	371	7	0,9620	6,56	9,35	2,80	0,103	0,207	–
6. PBMA + PS (1:1)	292	–	0,9510	4,84	7,00	2,96	0,087	0,173	–
	353	1	0,9920	4,65	6,30	1,65	0,059	1,118	–
7. PS + PC (1:1)	359	2	0,9110	4,12	6,10	1,98	0,072	0,143	–
	415	4	0,9530	5,24	10,23	4,99	0,104	0,209	–

components, the volume contracts more rapidly than in the pure component at the same temperature. This indicates that the process of microsegregation into two phases proceeds more intensively, in particular due to the initial loosening of the molecular packing, as follows from the expansion coefficients. Taking these data into account, we can once more consider the values of the constants of the modified SB equation.

According to the free-volume theory, the value of the SB constant is equivalent to the fractional free-volume at glass temperature. Thus the increase in the "constant" value can have only one meaning: an increase in the fraction of the free-volume of the component in the blend at the glass transition. It may be supposed that the microsegregation on heating leads to an increase in the density of each component (in good agreement with the gas chromatography method[95]). This in turn leads to the appearance of the excess free-volume in the interphase region. As in a micro-heterogeneous system, the ratio of the phase surface to its volume may be very large (in such a system colloid dispersion may occur). Thus we can conclude that in polymer blends some additional free-volume appears, which is localized at the interface of two phases. The distribution of the free-volume here is of a localized character, the main fraction of the free-volume being distributed between two phases in the interphase regions.

Let us introduce into Eq. (100) a coefficient which reflects the effective increase in the free-volume due to the appearance of an excess of interphase surface,

$$\frac{\Delta T_g}{\omega} = \beta \text{ const} \tag{101}$$

Values β are given in Table 4 for the universal value of the SB constant and for the values found experimentally for a given system. It is interesting to note that these data conflict with the results of calculating f_g for the same samples according to WLF theory.

We calculated the fractional free-volumes f_{g1} and f_{g2} according to the method developed by Covacs from the curves of the isothermal contraction[82]. The values obtained are not really the fractional free-volumes of the components at the corresponding temperatures, since in the calculation from contraction curves it is impossible to exclude the contributions of both components to the total free-volume. These contributions depend on the volume fraction of each component in the blend

Table 4. Values β for universal K_{un} and experimental K_{ex} constants of the Simha-Boyer equation

Blend	PBMA + PMMA	PBMA + PS	PS + PC
$\beta_1(K_{un})$	1.32	1.54	1.28
$\beta_1(K_{ex})$	1.48	1.73	1.52
$\beta_2(K_{un})$	1.85	1.05	1.87
$\beta_2(K_{ex})$	1.85	1.25	1.07

and on the difference $T_{g1} - T_{g2}$. If, however, we suppose that in accordance with WLF theory at the temperature $(T_g -50)$ the fraction of the free-volume is zero, then, taking into account that for all the systems the difference $T_{g1} - T_{g2} = 50°$, we can believe that the value f_{g1} may characterize the fractional free-volume of the component with the lower T_{g1}. Indeed, this value (Table 4) is very close to the universal value, whereas f_{g2} is higher than the universal value. It may be supposed that at T_{g2} the fraction of the free-volume of the whole system will be determined as

$$f_{g2}(T_{g2}) = w_1 [f_{g1} + \Delta\alpha(T_{g1} - T_{g2})] + w_2 f_{g2} \tag{102}$$

where $\Delta\alpha$ is the thermal expansion of f_{g1} in the interval between T_{g1} and T_{g2}. The calculated values correspond approximately to f_{g2} according to Covacs. Thus, we can see that in the microheterogeneous system the fractional free volumes of both components at T_g are not equal to each other nor to the universual value, and that therefore the iso-free-volume concept cannot be applied to such a system. This does not signify the inapplicability of the free-volume theory but only of the universal value f_g.

The activation energy of isothermal contraction in polymer blends calculated in [93] is considerably lower than for pure components, this pointing to the appearance of the free-volume as well, which facilitates the relaxation processes and diminishes the activation energy.

In [97] it was also shown on the basis of dilatometric data that the free-volume of PMMA in the mixture with polyvinylacetate increases with the increase in PVA concentration. In [98] a large difference was reported in the viscoelastic behavior of block copolymer from that predicted by WLF theory. This theory is believed to be useful only near the T_g of each component, not in the broad temperature interval including the transition from glassy to rubberlike state. This anomaly is thought to be connected with certain motions in the interphase regions, which should be looked upon as independent components of the mixture.

Another interpretation of the viscoelastic properties of polymer blends on the basis of the free-volume concept was given by Manabe and Takayanagi[99-101]. The viscoelastic properties of a system consisting of two individual phases were described on the basis of a mechanical model, proposed in [101]. This model consists of many elements with different glass temperatures and can be applied to the microheterogeneous systems. For such a system it is necessary to introduce the distribution function of T_g, which is supposed to be connected with the free-volume distribution function. It is assumed that the polymer blend may be divided into cells, their size being dependent on the physical properties under consideration. For glass temperature, the cell size may be connected with the size of the mobile unit (segment). The free-volume fraction f in the cell is different for various cells. The probability that the value f is between f and $f + df$ may be given by a function $F(f)df$, where $F(f)$ is the probability density and f is the distribution function of fractions of free-volume. The measurable fraction of the free-volume $(\bar{f})$ may be expressed as

$$\bar{f}(T) = \int_0^\infty f F(f) df \tag{103}$$

$$F(f) df = 1 \tag{104}$$

The dependence of the free-volume on the temperature for each cell is described by the equation

$$f = f_g + \Delta\alpha(T - T_g) U(T - T_g) \tag{105}$$

where $U(T - T_g)$ is the unit function equal to zero at $T < T_g$ and unity at $T > T_g$. From Eqs. (104) and (102) we have

$$\bar{f}(T) = \int_0^{T_g} \{f_g + \Delta\alpha(T - T_g)\} F(T_g) \, d\, T_g + \int_{T_g}^{\infty} f_g F(T_g) d\, T_g \tag{106}$$

From Eq. (106) it follows that

$$\{d^2 \bar{f}(T)/dT^2\} \, T = T_g = \Delta\alpha F(T_g) \tag{107}$$

The left side of Eq. (107) may be found experimentally. The probability that the glass temperature is situated between T_g and $T_g + d\,T_g$ is defined as $F(T_g) d\, T_g$ (distribution function of T_g). In this case the relationship between $F(T_g)$ and $F(f)$ is given as follows

$$F(T_g) d\, T_g \cong F(f) df \tag{108}$$

From Eqs. (105) and (108) it follows that

$$F(f) = F(T_g)/\Delta\alpha \tag{109}$$

It is worth noting that for two incompatible components the possibility of communization of the free-volume is supposed, when two components can be replaced by one new component. When the free-volumes of two components are communized, both components behave as one component with a common free volume after mixing. Depending on the mixture structure (compatible, incompatible, partially compatible, etc.) the shape of the free-volume distribution function will be different. According to Manabe[99], the free-volume of each component after mixing differs from that before mixing, *i.e.* the free-volume is dependent on the method of mixing.

The fraction of free volume of the i-th component (f_i) may be represented as

$$f_i = f_{i,0} + \sum_{j=1}^{n} a_i^j v_j + \sum_{j=1}^{n} \sum_{k=1}^{n} b_i^{jk} v_j v_k + \dots \tag{110}$$

$$v_i = 1 \tag{111}$$

where i, j, k and 0 denote i, j, k components in the mixture and pure component, v is the volume fraction, and a_i^j and b_i^j are coefficients reflecting the degree of intermolecular interaction and change in the energy of hole formation due to mixing.

The relationship between glass temperature and composition for different types of polymer blends may be established on the basis of Eq. (110). According to Hirai-Eiring theory[20], the partial free-volume is

$$f_i = A_i \exp(-U_{H_i}/RT) \tag{112}$$

where A_i is a constant and U_{H_i} the hole formation energy. For the mixture, U_{H_i} of each component changes due to changes in the intermolecular interaction. For this case

$$f_i = f_{i,0} \left\{ 1 - 1/RT \left(\sum_{j=1}^{n} (dU_{H_i}/d\,n_j)\,n_j + (1/2) \sum_{j=1}^{n} \sum_{k=1}^{n} \frac{d^2 U_{H_i}}{dn_i\,dn_k}\,n_i n_k + .. \right) \right\} \tag{113}$$

where n is a molar fraction of the component and f_0 the partial free-volume before mixing. It follows from the above discussion that a polymer mixture may be obtained in different states of mixing due to variations in the conditions of preparation. Thus, different distributions of free-volume are possible and there is no evidence that the glass temperature corresponds to the iso-free-volume state or has the same distribution function because of the various possible mixture structures. Thus, the different distribution functions for T_g correspond to different free-volume distribution functions and actual free-volume changes during mixing. Finally, the form of Eq. (109) does not correspond to the definition of free-volume according to Simha-Boyer since, if we take the average values f and T_g from the distribution function and replace $F(f)$ by $\bar{f}$ and $F(T_g)$ by $\bar{T}_g$, we obtain $\bar{f} = \bar{T}_g/\Delta\alpha$, i.e. T_g is divided and not multiplied by $\Delta\alpha$.

8. Dependence of the Fractional Free-Volume at Glass Temperature on Polymer Structure

We have already discussed the possibility of changes in fractional free-volume being related to the physical structure of polymers. To show this in greater detail, a special study was made[102]. The viscoelastic properties and relaxation time spectra were studied in a filled system where a powder of hardened epoxy resin was used as the filler and the same epoxy resin as the matrix. Thus the system was identical from the chemical point of view, the only difference being in the method of preparation. A three-dimensional network formed in the presence of the dispersed filler and therefore the process proceeded in the thin surface layer on the surface of the previously hardened resin (filler). The use of such a system enables the physical structure of the polymer to be changed without changing its chemical nature. At the same time, all the effects usually observed for filled polymers may be expected to be present.

From experimental data on viscoelastic properties the fractional free-volume was calculated according to the WLF equation for pure hardened resin, filled with different amounts of polymeric filler obtained from the same hardened resin. By using special methods for preparation of filled specimens, it was possible to obtain polymer material having filler properties similar to the properties of the surface layers of the epoxy matrix. In this way the attempt was made to increase artificially the proportion of polymer with properties changed under the influence of filler to reach the state of "pure surface layer". Calculations of fractional free-volume showed that these values differ for different compositions and do not correspond to the universal value although WLF theory was used as the basis for the calculation. Moreover, the calculated curves of the temperature dependence of the free-volume fraction above T_g have shown that the expansion coefficients of the free-volume also differ. In the case under consideration different physical structures were realized due to the formation of the polymer network in the surface layers the filler surface, as usually happens in filled systems. As is known[79], this induces considerable changes in the structure of the material. It is also possible that in these conditions a more defective network structure is formed. These results show that even the purely physical factors influencing the formation of the polymer network in the interface lead to such changes in the relaxation behavior and fractional free-volume that they cannot be described within the framework of the concept of the iso-free-volume state. It is of great importance that such a model has been devised for a polymer system that is heterogeneous yet chemically identical.

9. Conclusion

Everything discussed in the present paper shows that the free-volume concept, although very useful from the qualitative point of view, cannot be used for the quantitative description of many properties of polymer systems. This is especially clear when we consider glass-transition phenomena using the idea of the iso-free-volume state. Many experimental data, discussed above, show that this concept cannot be applied even to polymer materials having the same chemical nature but a different physical structure. From the experimental results Goldstein[104] had already concluded that the concept of free-volume cannot be correct. These conclusions were carefully discussed later[105].

We believe the difficulty is that the free-volume theory as applied to the glass-transition does not take account of the essential role of intra- and intermolecular interaction in the system and the flexibility of the polymer chains, all of which factors play an important role in the glass-transition phenomena.

We therefore believe one of the most pressing problems in the field of free-volume theory and glass-transition theory to be the development of new concepts and obtaining of new parameters for the corresponding states. Miller[37] as early as 1968 introduced the idea that the glass transition corresponds to the "iso-relaxation" state, which at molecular weights of polymers below the critical ones may be replaced by an "iso-viscous" state.

We believe that, as the very definition of free-volume characterizes the state of ordering in the system (especially for the cases when we consider the free-volume distribution), it may be preferable to apply the thermodynamic description of the processes at glass temperature.

The question of whether the glassy solidification is a purely kinetic process or may be considered as a thermodynamic transition has been frequently discussed[4, 105]. Gibbs and DiMarzio[106–108] have assumed that there will be a second-order transition at a temperature, T_2, at which the configurational entropy of the system becomes zero. Fewer conformations are available to the macromolecules at lower temperatures and as a result the molecular motion at T_2 is slower.

According to [1] the glass-transition occurs when the system reaches a certain minimum value of the configurational entropy ΔS_c at T_g, which can be calculated according to the equation

$$\Delta S_c = \Delta C \ln (T_g/T_0) \tag{114}$$

where ΔC is the change in heat capacity at glass temperature. Such an approach shows, in particular, that the values ΔS_c in unfilled and filled systems may be practically the same. Thus the idea that glass-transition phenomena represent an "iso-entropy" transition, according to the thermodynamic theory developed by Gibbs and DiMarzio, may be of great importance[54, 109]. It is interesting to note that, starting from completely different concepts, Adam and Gibbs[108] provide an alternative approach to the development of a WLF-type expression to account for the observed behavior of polymers close to the glass-transition temperature, and that these concepts do not rely on a free-volume assumption at all. This approach avoids the difficulties surrounding the free-volume figure of 0.025.

At the same time it is evident that the glass-transition phenomena are very complicated and that a different approach and different ideas will be necessary to explain them. This confirms the words written many years ago (in 1899) by the great Russian poet, Valery Briusov, that "there are many truths and they often contradict one another".

10. References

[1] Adam, G., Gibbs, J.: J. Chem. Phys. 43, 139 (1965)
[2] Goldstein, M.: J. Chem. Phys. 39, 3369 (1963)
[3] Haward, R., in: The physics of glassy polymers. Haward, R. H. (ed.) London: Applied Science Publ. Ltd., p. 1–53
[4] Roberts, G., White, E., in: Physics of glassy polymers, p. 153
[5] Roger, S. Mandelkern, L.: J. Phys. Chem. 61, 985 (1957)
[6] Covacs, A.: Rheol. Acta 5, 262 (1966)
[7] Ono, S., Kondo, S.: Molecular theory of surface tension in liquids, 1960
[8] Siow, K., Patterson, D.: J. Phys. Chem. 77, 356 (1973)
[9] Hill, T.: Statistical Mechanics. New York: McGraw-Hill 1956
[10] Hildebrand, J.: Phys. Rev. 34, 984 (1929)
[11] Haward, R.: J. Macromol. Sci.-Revs. Macromol. Chem. C4, 191 (1970)
[12] Ferry, J.: Viscoelastic properties of polymers
[13] Doolittle, A.: J. Appl. Phys. 22, 1471 (1951)
[14] Fujita, H.: Fortschr. Hochpolymer Forsch. 3, 1 (1961)
[15] Machin, D., Rogers, C.: Makromol. Chem. 155, 269 (1972)
[16] Waghizadeh, J., Ueberreiter, K.: Koll. Z. Z. Polymere 250, 98 (1972)

[17] Patterson, D.: Macromolecules *2*, 672 (1969)

[18] Bueche, F.: J. Chem. Phys. *21*, 1850 (1953)

[19] Fox, T., Flory, P.: J. Appl. Phys. *21*, 581 (1950)

[20] Hirai, N., Eiring, H.: J. Polymer Sci. *37*, 51 (1959)

[21] Wunderlich, B.: J. Phys. Chem. *64*, 1052 (1960)

[22] Simha, R., Boyer, R. J.: Chem. Phys. *37*, 1003 (1962)

[23] Williams, M., Landell, R., Ferry, J.: J. Amer. Chem. Soc. *77*, 3701 (1955)

[24] Turnbull, D., Cohen, M.: J. Chem. Phys. *34*, 120 (1961)

[25] Boyer, R.: Rubber Chem. Techn. *36*, 1303 (1963)

[26] Miller, A.: J. Polymer Sci. *A 2*, 1095 (1964)

[27] Litt, M., Tobolsky, A. J.: Macromol. Sci.-Phys. *B 1*, 433 (1967)

[28] Cohen, H., Turnbull, D.: J. Chem. Phys. *31*, 1164 (1959)

[29] Miller, A.: Macromolecules *2*, 355 (1969)

[30] Vogel, H.: Physik Z. *22*, 645 (1921)

[31] Miller, A.: J. Phys. Chem. *67*, 103 (1963)

[32] Hirai, N., Eiring, H.: Appl. Phys. *29*, 810 (1958)

[33] Hirai, N., Eyiring, H.: J. Polymer Sci. *37*, 51 (1959)

[34] Smith, R.: J. Polymer Sci. *A 2*, 8, 1337 (1970)

[35] Sanditov, D., Bartenev, G.: Phisika i chimija stekla *1*, 414 (1975)

[36] Miller, A.: Macromolecules *5*, 674 (1970)

[37] Miller, A.: J. Chem. Phys. *49*, 1393 (1968)

[38] Bueche, F.: J. Chem. Phys. *24*, 418 (1956)

[39] Miller, A.: J. Polymer Sci. *A 2*, 6, 249 (1967)

[40] Miller, A.: Amer. Chem. Soc. Polymer Prepr. *15*, 412 (1974)

[41] Miller, A.: J. Polymer Sci. *A 2*, 4, 415 (1966)

[42] Kanig, G.: Kolloid Z. *190*, 1 (1963)

[43] Frenkel, J.: Kinetic theory of liquids, Akad. Nauk SSSR, Moscow, 1943

[44] Kanig, G.: Kolloid Z. *203*, 161 (1965)

[45] Kanig, G.: Kolloid Z. *233*, 829 (1969)

[46] Lipatov, Yu.: Vysokomolek. soed. *B 10*, 527 (1968)

[47] Lipatov, Yu., Geller, T.: Vysokomolek. soed. *A 9*, 223 (1967)

[48] Kozlov, P., Timofeeva, V., Kargin, V.: Dokl. Akad. Nauk SSSR *148*, 886 (1963)

[49] Sanchez, J.: J. Appl. Phys. *45*, 4204 (1974)

[50] Sharma, S., Mandelkern, L., Stehling, F.: J. Polymer Sci. *B 10*, 345 (1972)

[51] Boyer, R., Simha, R.: J. Polymer Sci. *B 11*, 33 (1973)

[52] Simha, R., Weil, C.: Macromol. Sci.-Phys. *B 4(1)*, 215 (1970)

[53] Moacanin, J., Simha, R.: J. Chem. Phys. *45*, 964 (1966)

[54] Gibbs, J., DiMarzio, E.: J. Chem. Phys. *28*, 373 (1958)

[55] Ellerstein, S.: J. Polymer Sci. *B 1*, 311 (1963)

[56] Kästner, S.: J. Polymer Sci. *C, N 16*, 4121 (1968)

[57] Hoffman, J., Williams, G., Passaglia, E., in: Transition and relaxation in Polymers. Boyer, R. (ed.). John Wiley 1966

[58] Williams, G.: Trans. Faraday Soc. *60*, 1556

[59] Simha, R.: J. Macromol. Sci. *B 5*, 331 (1971)

[60] Letunovsky, M., Minkin, E.: Zelenev Yu.: Vysokomolek. soed *A 15*, 345 (1973)

[61] Bunijat-Zade, A., Ismaylov, T.: Plast. massy *N 4*, 70 (1974)

[62] Curro, I.: J. Macromol. Sci. *C 11*, 321 (1974)

[63] Matsuoko, S., Yshida, Y., in: Transition and relaxation in polymers, Boyer, R. (ed.), Interscience Publ. 1966

[64] Miller, A.: J. Polymer Sci. *A 2.6*, 249 (1968)

[65] Beevers, R.: J. Polymer Sci. *A 2*, 5257 (1964)

[66] Krause, S.: J. Polymer Sci. *A 3*, 3573 (1965)

[67] Lipatov, Yu., Privalko, V.: J. Macromol. Sci.-Phys. *B 7(3)*, 431 (1973)

[68] Kurata, M., Stockmayer, W.: Fortschr. Hochpolymer. Forsch. *3*, 19 (1963)

[69] Privalko, V., Lipatov, Yu.: Vysokomolek. Soed. *A 13*, 2733 (1971)

[70] Litt, M.: Amer. Chem. Soc. Polymer Prepr. *14*, 109 (1973)

[71] Privalko, V., Lipatov, Yu.: J. Macromol. Sci.-Phys. *B 9(3)*, 551 (1974)

[72] Kästner, S.: Kolloid Z. Z. Polymere *206*, 143 (1965)

[73] Bondi, A.: J. Phys. Chem. *58*, 929 (1954)

[74] Sanditov, D., Bartenev, G.: J. Phys. Chimii *46*, 2214 (1972)

[75] Kilian, H.: Koll. Z. Z. Polymere *252*, 353 (1974)

[76] Rasumovskaya, I., Bartenev, G., in: Glasslike state (Russ.) Nauka, 1971, p. 34

[77] Mason, P.: Polymer *5*, 625 (1964)

[78] Lipatov, Yu.: Pure a. Appl. Chem. *43*, 273 (1975)

[79] Lipatov, Yu., Sergeeva, L.: Adsorption of polymers. John Wiley 1974

[80] Lipatov, Yu., Moysja, E., Semenovich, G.: Polymer *16*, 582 (1975)

[81] Lipatov, Yu.: Adv. Polymer Sci. *22*, 1 (1977)

[82] Lipatov, Yu., Geller, T.: Vysokomolek. soed. *8*, 593 (1966)

[83] Lipatov, Yu., Babich, V., Rosovizky, V., in: Physical chemistry of polymer compositions (Russ.). Kiev 1974, p. 32

[84] Lipatov, Yu., Babich, V., Korzguk, N.: Vysokomolek. soed. *A 16*, 1629 (1974)

[85] Lipatov, Yu., Privalko, V., Shumsky, V.: Vysokomolek. soed. *A 15*, 571 (1973)

[86] Krauss, G., Gruver, J.: J. Polymer Sci. *A 2.8*, 571 (1970)

[87] Lipatov, Yu., Privalko, V.: et al. Vysokomolek. soed. *A 19*, 1756 (1977)

[88] Lipatov, Yu.: J. Polymer Sci. *C, N 42*, 855 (1973)

[89] Block-polymers. Aggarwal, S. (ed.). Plenum Press 1970

[90] Lipatov, Yu., Kercha, Yu.: Vysokomolek. soed. *A 15*, 1057 (1973)

[91] Kuleznev, V., in: Mnogokomponentnie polimernie sistemy. Chimija: Moscow 1974, p. 10

[92] Lipatov, Yu., Fabuljak, F.: Dokl Akad. Nauk SSSR *205*, 635 (1972)

[93] Lipatov, Yu., Vilensky, V.: Vysokomolek. soed. *A 17*, 2069 (1975)

[94] Manabe, S., Takayanagi, M.: J. Chemà Soc. Japan, Ind. Chem. Sec. *23*, 57 (1967)

[95] Nesterov, A., Lipatov, Yu.: Dokl. Akad. Nauk SSSR *222*, 609 (1975)

[96] Covacs, A.: J. Polymer Sci. *30*, 131 (1958)

[97] Uchida, Y.: Bull. Tokyo Inst. Techn. *N 76*, 45 (1966)

[98] Shen, M., Kaelble: J. Polymer Sci. *B 8*, 149 (1970)

[99] Manabe, S., Murakami, K., Takayanagi, M.: Memoirs of the Fac. Eng. Kyuchu Univ. *28*, 295 (1969)

[100] Manabe, S., Takayanagi, M.: Kogyo Kagaku Zasshi *70*, 525 (1970)

[101] Takayanagi, M., Hazima, H., Ivata, Y.: Mem. Fac. Eng. Kyushu Univ. *23*, 57 (1963)

[102] Lipatov, Yu., Babich, V., Rosovizky, V.: Europ. Polymer J. *13*, 651 (1977)

[103] Lipatov, Yu., Babich, V., Rosovizky, V.: J. Appl. Polymer Sci. *20*, 1787 (1976)

[104] Goldstein, M.: J. Chem. Phys. *39*, 3369 (1963)

[105] Rehage, G., Borchard, H., in: The physics of glassy polymers, p. 54

[106] Gibbs, J.: J. Chem. Phys. *25*, 185 (1965)

[107] Gibbs, J., DiMarzio, E.: J. Polymer Sci. *A 1*, 1417 (1963)

[108] Gibbs, J., DiMarzio, E.: J. Res. Nat. Bur. Stand. *68 A*, 611 (1964)

[109] DiMarzio, E.: J. Appl. Phys. *45*, 4143 (1974)

Received February 21, 1977

Model Networks

Jean E. Herz and Paul Rempp
Centre de Recherches sur les Macromolécules – C.N.R.S. – 6, rue Boussingault,
67083 Strasbourg Cédex, France

Werner Borchard
Physikalisch-Chemisches Institut der Technischen Universität Clausthal, A.-Römer-Straße 2 A,
3392 Clausthal-Zellerfeld, West Germany

Model networks, synthesized by endlinking processes, contain few structural defects and are close to ideality. Spring-suspended bead models seem to fit adequately with the structural data obtained on labelled model networks and with the swelling and uniaxial deformation behavior of these networks. (67 refs.)

Table of Contents

1. Introduction

The methods generally used to synthesize polymeric networks cannot provide for
any precise control, nor even for any knowledge of their internal structure.

Radical copolymerization[1, 2], for instance, is basically a random process. Even
if the percentual consumption of both monomers proceeded at the same rate — which
is generally not the case — the average length of the elastic chains could only be
evaluated roughly, and the distribution of the chain lengths is bound to be very
broad. Moreover, the reaction medium gels at an early stage of the reaction and the
gel point is therefore not well defined, as it cannot be related to segment concentration, chain length, end-to-end distance, and functionality. Various chain transfer
processes are responsible for an increase of the number of pendant chains. In some
cases the gel is not even macroscopically homogeneous, particularly when syneresis[3, 4]
— solvent expulsion — takes place during the process.

Similar criticism can be raised against networks synthesized by polycondensation
processes involving one tri- or tetrafunctional component to achieve crosslinking.

Networks are also often obtained by vulcanization processes[5, 6] carried out on
linear "primary" macromolecules. Crosslinking occurs at random, and the length of
the elastically effective chains varies within large limits for a given sample. A precise
knowledge of the total number of branch points formed is impossible. The number
of pendant chains can be estimated roughly as twice the number of primary chains.
Loops and double connections are likely to occur during the vulcanization reaction,
both of them reducing the number of elastically effective network chains.

Such networks have been widely used to establish whether the theories of rubber
elasticity and of equilibrium swelling are valid. But these theories are based on a
number of hypotheses which are obviously far from being fulfilled by the above networks. The so-called ideal networks should obey the following requirements:

1. They should be homogeneous: this does not mean only that syneresis should
be avoided, it also implies that the segment density and the crosslink density should
both remain constant throughout the sample.

2. They should consist of elastically effective chains only. An elastically effective
chain should connect two different crosslinks, and two such crosslinks should be
tied by only one elastic chain. This means that the gel should contain no defects
such as pendant chains (one end of which only is connected with a crosslink), loops
(chains linked at both ends to the same crosslink), or double connections. Physical
crosslinks (permanent entanglements) should be prohibited, too.

3. The elastically effective network chains should obey Gaussian statistics. They
should therefore be long enough, and their average degree of polymerization should
be known. In addition the distribution of chain-lengths is expected to be rather
narrow.

4. The functionality of the crosslinks should be known, and constant throughout
the gel. The functionality is the number of elastically effective network chains which
are tied to one given crosslink.

2. Synthesis of Model Networks

New methods are being developed in various laboratories to synthesize well-defined networks exhibiting structures as close as possible to ideality. The principle of these so-called endlinking methods is to separate the polymerization process from the network-forming reaction. The first step aims at preparing a linear precursor polymer, fitted at both ends with reactive groups. In the second step bonds are established between several precursor chain ends to form the crosslinks. The methods which were used to synthesize so-called model networks have already been described, and we shall only summarize them here:

2.1. Anionic Block Copolymerization

The first method, which is used in many instances, proceeds by anionic block co-polymerization. In a first step styrene is anionically polymerized using an efficient bifunctional initiator. The polymer obtained has a well-defined molecular weight, it exhibits a rather narrow molecular weight distribution and it has carbanionic living sites at both ends of its linear chains[7].

These sites can be used to initiate the polymerization of a small amount of divinylbenzene (DVB)[8-10]. The bifunctional monomer will polymerize to small tightly crosslinked nodules; each of them is connected with the f chain ends which have participated in its initiation process. The average functionality f of the nodules is not directly accessible. However, it was shown in the case of star-shaped molecules that f increases with the overall concentration of the precursor, and, to a smaller extent, with the amount of DVB added per living end. f is almost independent of the precursor chain length[11].

This method was applied to synthesize various networks, with elastic chains of different nature: polystyrene, polymethacrylates, polyvinylpyridine, and more recently polydienes. In some cases ethylene dimethacrylate is used to achieve cross-linking[9], because of its higher electrophilicity.

This method was also used to synthesize gels with labelled crosslinks, designed for low-angle X-ray scattering experiments[12], (the nodules contain heavy atoms such as iron) or for neutron coherent scattering measurements[13], whereby the nodules are surrounded by deuterated segments to provide for adequate contrast.

2.2. Anionic Deactivation Crosslinking

A second method, using also anionic polymerization techniques was developed to achieve a better knowledge of the functionality of the branch points. In this method the bifunctional "living" polymer species is prepared under the same conditions as above. It is then reacted with a stoichiometric amount of a plurifunctional electro-philic reagent[14, 15]. The chief difficulty is to find a reagent the electrophilic functions of which are isoreactive, to ensure that the reaction will go to completion.

Adequate mixing has also to be provided for. Upon reaction, the polymer solution gels, and it can be expected that the precursor chains are converted into elastic chains, whereas the crosslinks should exhibit the functionality of the electrophilic reagent[15, 16]. Suitable electrophilic reagents are: tris(allyloxy)s-triazine(trifunctional) and a similar tetrafunctional compound. Plurifunctional isocyanates have also been used successfully for such syntheses[17].

2.3. Endlinking of Telechelic Polymers

The third method consists in using α, ω-bifunctional polymers of known molecular weight, and in reacting them with adequate plurifunctional reagents[17–22]. In some cases the α, ω-bifunctional precursor polymer is also obtained anionically – this provides for a sharp molecular weight distribution – in other cases it is obtained by other methods.

A precise characterization of the telechelic precursor is necessary, prior to the crosslinking reaction, since precise stoichiometry must be achieved if the number of defects in the network is to be kept as small as possible.

An alternate way has sometimes been chosen[23, 24]. It consists in preparing star-shaped macromolecules with functional groups at the end of all branches. In a second step coupling is achieved by means of a bifunctional reagent.

3. Ideality of the Model Networks

The question now arises whether the so-called model networks are really close to ideality, and whether they may be used to check the validity of the existing theories of equilibrium swelling and of rubber elasticity. This question deserves some discussion.

Since the method proceeds by endlinking of preexisting linear precursor chains – whereby they are converted into elastic chains – an adequate control of the average length of these chains and of the corresponding polydispersity becomes possible. The structure of the network – and such properties as the swelling degree which depends upon the length of the elastic chains – can be chosen in advance. Networks with well-defined properties can thus be synthesized at will.

The networks obtained are homogeneous, exhibiting constant segment densities and constant crosslink densities throughout a sample.

It has to be pointed out that the segment concentration hardly changes upon crosslinking. This segment concentration v_c may be chosen as the reference state, because upon crosslinking the conformational changes of the chains are minimized at v_c.

However, the presence of defects such as loops, pendant chains, double connections, cannot be excluded. The former are more likely to occur when the polymer concentration of the reaction medium is low, the latter when it is very high. The presence of loops and of pendant chains reduces the number of elastically effective network chains.

Conversely, permanent chain entanglements[25] act as additional, but not fixed, crosslinks, which are more likely to form when the functionality of the crosslinks is rather high, as it will be seen in a later section of this review.

4. Structure of Model Networks

4.1. Experimental Results

We shall consider now a number of results on the intimate structure of model networks.

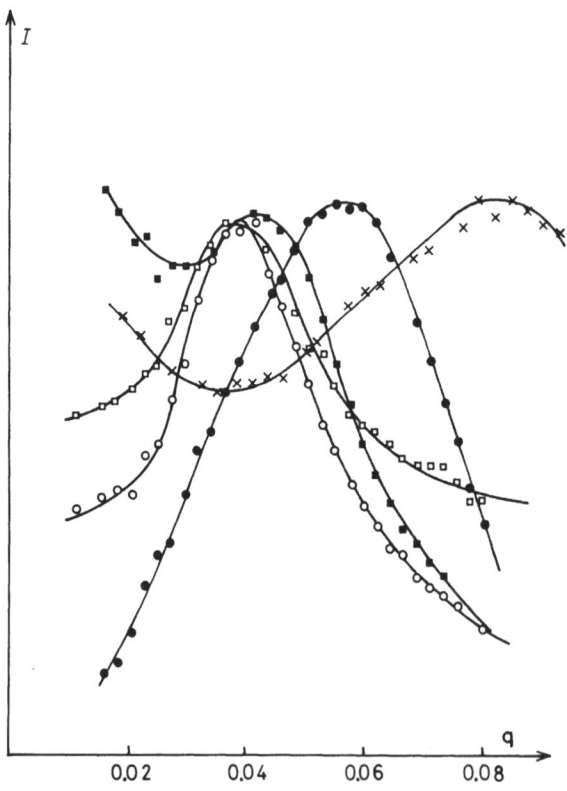

Fig. 1. Example of neutron scattering envelope of PS/DVB networks with deuterated crosslinks swollen in various solvents.

I: relative scattering intensity, $q = \dfrac{4\pi}{\lambda} \sin \dfrac{\theta}{2}$ (in A^{-1}); 1 A = 0.1 nm

x: dry network; ○ CS_2; □ benzene; ■ tetrahydrofurfurylic alcohol (THFA); ●: cyclohexane

If networks with iron-labelled crosslinks are studied by small-angle X-ray scattering, the diagrams obtained exhibit one single, rather broad, but well-defined diffraction band. For a given network the angular position of the maximum of the peak is a function of the degree of swelling of the gel. For a series of homologous networks – differing merely in the average length of their elastically effective chains – the peak is shifted to lower scattering angles as the mean molecular weight of the network chains increases[12].

Similar results have been obtained from neutron coherent scattering experiments[13] on networks with deuterium-labelled crosslinks. Here again a single well-defined scattering maximum is observed, the angular position of which depends on the length of the network chains and on the swelling ratio of the gel (Fig. 1).

Moreover, these results can even be obtained on unlabelled networks, when the swelling solvent is deuterated[26, 27]. Use is made of the fact that in the vicinity of the crosslinks the segment density is higher than elsewhere; consequently the number of deuterated solvent molecules is less in the vicinity of the crosslinks. The contrast thus produced between deuterated and undeuterated domains is sufficient to provide for a single well-defined peak, the angular position of which is quite compatible with that obtained on labelled networks of the same characteristics in the nondeuterated solvent; the activity of the two solvents – deuterated and undeuterated – with respect to the polymer systems investigated is almost the same[28].

It follows from these small-angle X-ray scattering experiments and from these neutron coherent scattering determinations that there is a well-defined correlation distance between first neighbor crosslinks in such model networks[12, 13]. This means that the distribution of distances between neighbor crosslinks is rather narrow. There is, however, no correlation between the position of second neighbor nodules: the gels behave in that respect as amorphous compounds or as liquids.

The average distance d between neighboring nodules can be calculated using Bragg's law:

$$d = \frac{2\pi}{\lambda} \sin \frac{\theta}{2}$$

It has been found that for a homologous series of model networks the value of d increases as the average length of the elastic chains increases. For a given network d is found to increase as the degree of swelling becomes larger.

More precisely, neutron scattering experiments showed that the product $d\,Q^{-1/3}$ is a constant for a given network[13] – Q being the equilibrium swelling degree in the swelling solvent used, defined as the ratio of the total volume of the gel to the volume of the solvent free network. Incidently, this is a very convincing argument in favor of the affine character of the deformation of a network induced by swelling, at least for degrees of swelling not exceeding 10. This result also illustrates the homogeneity and the isotropy of the model networks investigated.

However, the results of small-angle X-ray scattering experiments on iron-labelled networks have not entirely confirmed the above results: for a given network the product $d\,Q^{-1/3}$ is not found to be a constant. The reason for this discrepancy is yet unknown.

4.2. Proposed Model for the Network Structure

The results obtained by small-angle X-ray scattering and by neutron coherent scattering can give us some insight into the internal structure of our model networks: it can be expected that such a swollen network should behave like an ensemble of spring-suspended beads. The crosslinks are the beads and the elastically effective network chains are the springs[a] connecting them[14, 27, 29]. It is obviously not required, in such a system, that all first neighbor crosslinks be connected by elastic chains, nor that elastic chains solely connect first neighbor crosslinks. To be more precise, we have to distinguish between *chemical functionality* – which is the average number of elastically effective network chains connected with one given nodule – and *geometric functionality* – which is the number of crosslinks which are first neighbors to one given nodule in the network. If the chemical functionality of the crosslinks is low (3 to 5), the probability for network chains to connect nodules which are not first neighbors can be expected to be low also. But if the functionality is high nothing should prevent some elastic chains connecting nodules which are further apart. Another way of handling the same problem is to use the connectivity factor x^2, first introduced by Ziabicki[30], which is defined as the ratio of the mean square end-to-end distance of the elastic chains to the square of the average distance between two first neighbor crosslinks:

$$x^2 = \frac{\langle r_d^2 \rangle}{\langle d^2 \rangle} \tag{1}$$

If the elastic chains solely connect first neighbor nodules, $x^2 = 1$. If the proportion of chains linking crosslinks which are not first neighbors increases, the value of x^2 also increases. The parameter x^2 therefore characterizes the degree of interpenetration of the network chains, and it should be related to the proportion of entanglements present in the network. In practice, it can be expected that for networks in which the functionality of the crosslinks is low, entanglements are quite unlikely, and the value of x^2 should stay close to unity.

It has to be emphasized that x^2 is an intrinsic parameter of the network. Its value should be the same in the unswollen state and in the presence of *any* swelling solvent.

4.3. Radius of Gyration of Elastic Chains

Neutron scattering can also be used to determine the radius of gyration of a polymer chain: the angular variation of the scattering intensity (at low scattering angles) leads to the value of the radius of gyration ρ of deuterated chains in a nondeuterated matrix[31]. To achieve this type of measurement on individual elastic chains of a gel,

---

[a]   It should be remembered, however, that the springs connecting the beads are entropy springs, and that the network chains are flexible either because they are above their glass transition temperature, or because they are imbedded in a solvent.

networks had to be synthesized on purpose, with a small proportion of their elastic chains entirely deuterated[13, 26].

If such a network with labelled chains is studied in the presence of various swelling solvents, the product $\rho Q^{-1/3}$ is found to vary far beyond experimental error, contrary to the product $d Q^{-1/3}$, discussed above. This result was however expected: if Gaussian statistics is applied to a chain the ends of which are maintained at a distance d, the radius of gyration of that chain can be expressed as[32]:

$$\rho^2 = \frac{1}{12} (N b^2 + d^2) \tag{2}$$

where N is the number of segments and b their length. If the chain is free d^2 is equal to $N b^2$, and the well-known expression results:

$$\rho^2 = \frac{1}{6} N b^2$$

If we consider a swollen gel, swelling results in a displacement of the chain ends, d^2 becomes larger than $N b^2$, and the proportionality between radius of gyration and mean square end-to-end distance is no longer valid. One cannot therefore expect the product of $\rho Q^{-1/3}$ to remain constant. However, the end-to-end pulling effect of swelling is not sufficient to account for the values of ρ which are in all cases too low to fit the expectations. The highest discrepancies are observed for networks swollen in very good solvents[13], which are those for which the end-to-end pulling effect on the elastic chains is the largest.

It follows that the "nascent"[b] networks are best fitted for defining a reference state, because at the swelling ratio Q_c the elastic chains are supposed to have undergone the smallest conformational changes with respect to the free chains they were before crosslinking.

4.4. Memory Term

First introduced as "front factor" by Tobolsky[33], and defined, according to Dusek and Prins[3], as

$$h^{2/3} = \frac{\langle r_d^2 \rangle}{\langle r_{os}^2 \rangle} \tag{3}$$

the "memory term" deserves some discussion. $\langle r_d^2 \rangle$ is the mean square end-to-end distance of the network chains in the dry network, and $\langle r_{os}^2 \rangle$ is the mean square end-to-end distance of the corresponding free chains (prior to crosslinking) in the reference state.

---

[b] A "nascent" network is a network in the condition in which it was formed: its swelling ratio Q_c is related to the polymer concentration v_c by: $Q_c = v_c^{-1}$.

The choice of the swollen reference state still remains somewhat ambiguous. It is tempting to consider the state with swelling degree Q_c of the nascent network as the reference state, since the gel point has a physical meaning, at least for networks synthesized by endlinking processes: no significant change of the segment concentration occurs upon crosslinking. If we call $\langle r_c^2 \rangle$ the mean square end-to-end distance of the elastic chains in the nascent network, characterized by a swelling ratio Q_c, we may write:

$$h^{2/3} = \frac{\langle r_d^2 \rangle}{\langle r_c^2 \rangle} \cdot \frac{\langle r_c^2 \rangle}{\langle r_{os}^2 \rangle} = \frac{1}{Q_c^{2/3}} \cdot \frac{\langle r_c^2 \rangle}{\langle r_{os}^2 \rangle} \tag{4}$$

The ratio $\langle r_c^2 \rangle / \langle r_{os}^2 \rangle$ characterizes the effect of crosslinking upon the chain dimensions. If the crosslinking process does not involve large chain extensions or contractions, the value of this ratio should remain close to unity.
$h^{2/3}$ should then be related primarily to the concentration at which crosslinking took place. It should be independent of the nature of the swelling solvent and of the equilibrium swelling degree.

Whether this choice of the relaxation state is justified is still debatable. However, an argument in favor of this choice is given by the fact that the concentration v_c at which crosslinking ocurred has a marked influence upon the properties of the network[22, 34].

Another possible choice for the reference state is the equilibrium swelling degree of the network. In that case the thermodynamic characteristics of the actual polymer-swelling solvent system would be taken into account. However this choice does not consider the conditions of network formation. Moreover when the gel is swollen to equilibrium, the elastic chains are extended with respect to the corresponding free chains, and the extension ratio is not easy to evaluate[13].

Attention should be drawn to the fact that if the state with the swelling ratio $Q_c (= v_c^{-1})$ of the nascent network is chosen as the reference state, the memory term $h^{2/3}$ should be independent of the molecular weight of the elastic chains. This point is still somewhat controversial, and though experimental data support this statement in several cases[35, 36] there are other cases in which the opposite was observed[14, 22].

4.5. Intercrosslink Distances

Another approach to this problem arises from the fact that the number of crosslinks in a network can be evaluated directly. The overall concentration of polymer in the swollen network being v (the reciprocal of the volume swelling degree Q), the number of elastic chains per unit volume of swollen network is given by:

$$n = \frac{\rho_2 \cdot v \mathcal{N}}{M} \tag{5}$$

where M is the molecular weight of the elastic chains, ρ_2 is the density of the dry network and $\mathcal{N}$ Avogadro's number. If f is the chemical functionality of the crosslinks, the number of such crosslinks, under the hypothesis of ideality is given by:

$$n_* = \frac{2}{f} \cdot \frac{\rho_2 \cdot \mathcal{N} v}{M} \tag{6}$$

Therefore the average volume available for each crosslink in the swollen gel can be expressed as:

$$V_* = \frac{fMQ}{2\rho_2 \mathcal{N}} \tag{7}$$

To evaluate the average distance between neighbor crosslinks in the swollen network, the geometric functionality – i.e., the number of nearest neighbors to one given crosslink – has to be taken into consideration by a factor C:

$$d = C V_*^{1/3} \tag{8}$$

C is equal to unity when each nodule has six first neighbors, by analogy with a simple cubic lattice[29]. Similarly C is of the order of 0.87, 1.09, 1.12 when the geometric functionality is of the order of 4, 8, 12, respectively; these values originate from calculations carried out on diamond-type, centered, and face-centered cubic lattices, which exhibit precisely these geometric functionalities (or coordination indices). In any case, C is a constant for a given network, and its value is never very far from unity.

Since swelling introduces an affine deformation of the network, the above calculation holds for any swelling degree and the distance d between first neighbor crosslinks can be written as:

$$d = C \left[\frac{fMQ}{2\rho_2 \mathcal{N}} \right]^{1/3} \tag{9}$$

This relation should also hold for unswollen networks ($Q = 1$), provided no voids are formed upon drying.

This expression can be used to evaluate the intercrosslink distance d_c in the nascent gel, whereby $Q_c = v_c^{-1}$ is the swelling degree of the gel upon its formation. However the intercrosslink distance d can only be considered identical with $\langle r^2 \rangle^{1/2}$ – the root mean square end-to-end distance of the elastic chains – if the overwhelming majority of the elastic chains connect first neighbor crosslinks, i.e., when the elastic chains hardly interpenetrate.

Let us consider now the above equation in more detail: (Table 1)

1. For a given network the intercrosslink distance d should be proportional to $Q^{1/3}$. This is exactly what was found by neutron scattering on networks with deuterium-labelled nodules. In particular d_c is the distance between neighbor crosslinks in the nascent network, the swelling ratio being Q_c (Fig. 2).

2. For a homologous series of networks, assuming that f is the same over the whole series, the intercrosslink distance d should obviously be a linear function of $(MQ)^{1/3}$. For unswollen networks ($Q = 1$), d should be proportional to $M^{1/3}$. Ex-

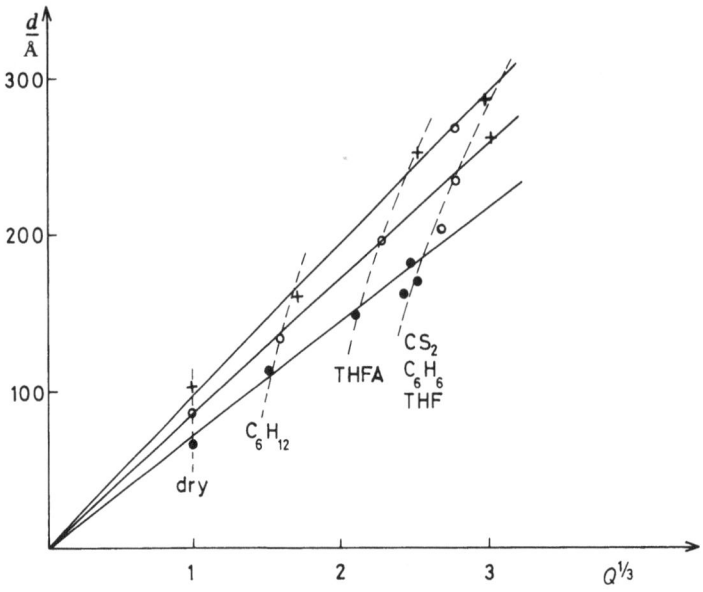

Fig. 2. Average correlation distances d between deuterated crosslinks of PS/DVB networks swollen to equilibrium vs $Q^{1/3}$.

●: $M = 22,800$; ○: $M = 35,500$; +: $M = 46,400$ (in g · mol^{-1})

Table 1. Experimental results of neutron scattering measurements on PS networks with deuterated crosslinks

Solvent	Q	$M = 21,000$ $\dfrac{d}{Å}$	$dQ^{-1/3}$	Q	$M = 30,000$ $\dfrac{d}{Å}$	$dQ^{-1/3}$
Dry	1	95.4	–	1	112.2	–
C_6H_{12}	2.34	136.3	102.5	2.67	162.3	116.8
THFA	5.1	178	103.5	6.3	227	122.7
C_6H_6	9.8	209.8	98	12.6	252.5	108.4

Solvent	Q	$M = 44,000$ $d(Å)$	$dQ^{-1/3}$	Q	$M = 50,000$ $d(Å)$	$dQ^{-1/3}$
Dry	1	130.5	–	1	136.6	–
C_6H_{12}	2.9	192.8	134.8	3.01	217.2	150.8
THFA	8.1	272.1	135.8	9.01	307.4	147.8
C_6H_6	17.5	298.9	115	19.5	329.4	122.5

Q = swelling ratio.
d = average correlation distance between crosslinks in Å.
THFA = tetrahydrofurfurylic alcohol.
M in g · mol^{-1}.

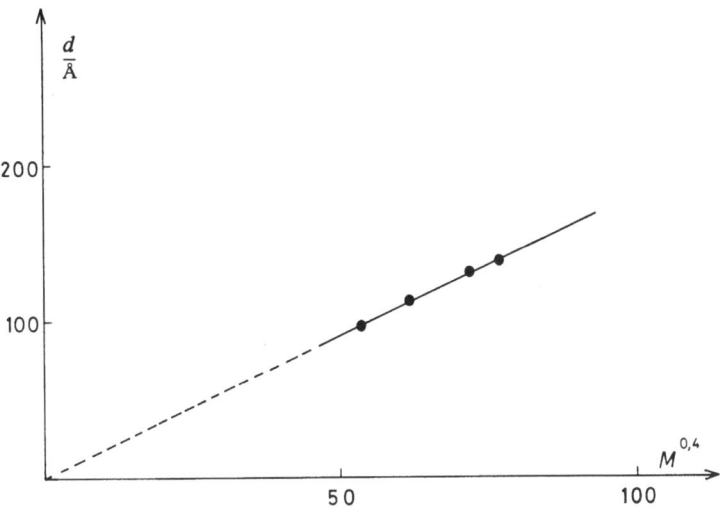

Fig. 3. Average correlation distances d between deuterated crosslinks of PS/DVB networks vs $M^{0.4}$: dry networks (M in g · mol^{-1})

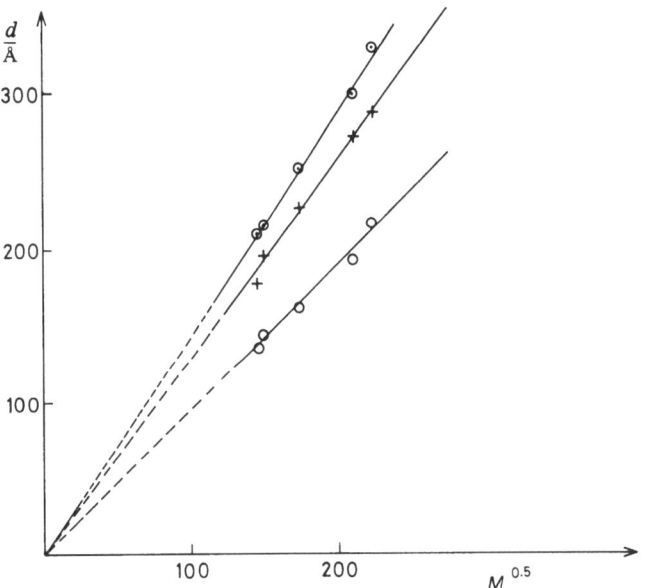

Fig. 4. Average correlation distances d between crosslinks vs $M^{1/2}$ in PS/DVB networks swollen to equilibrium.
⊙: benzene; +: tetrahydrofurfurylic alcohol (THFA); ○: cyclohexane

perimentally the exponent is found to be somewhat larger (0.4). This may be due to formation of nonequilibrium states (voids) upon drying, since the polystyrene networks are being studied far below their glass transition temperature (Fig. 3).

A linear variation of d versus $M^{1/3}$ should also be expected for polystyrene networks in cyclohexane at 20 °C, since under these conditions Q is almost independent of M[37]. Experimentally, X-ray diffraction measurements confirm that d is proportional to $M^{1/3}$ for such systems[12], whereas neutron scattering yielded exponents somewhat higher (0.4).

For homologous networks swollen in "good" solvents, in which case Q is roughly proportional to $M^{3/5}$, d is expected to be proportional to $M^{8/15}$. The values of d originating from neutron scattering determinations are found to be proportional to $M^{1/2}$ for homologous networks[13]. If account is taken to the rather low accuracy of these measurements in good solvents, one can consider $1/2$ to be rather close to the expected $8/15$ (Fig. 4).

5. Swelling Behavior of Model Networks

5.1. Some Theoretical Considerations

Let us now consider in more detail the swelling behavior of the model networks. It is well-known that the difference between the chemical potential of the solvent in the gel μ_1, and that of the pure solvent μ_{01} comprises two terms[1, 3]: one dilution term and one characterizing the elastic deformation of the gel.

$$\mu_1 - \mu_{01} = RT[\ln(1 - v) + v + \chi v^2 + V_{01}v(A\ v^{1/3}h^{2/3} - B\ v)] \tag{10}$$

At swelling equilibrium $(\mu_1 - \mu_{01})$ is zero, and therefore:

$$v = \frac{\ln(1 - v) + v + \chi v^2}{V_{01}(B\ v - A\ v^{1/3}h^{2/3})} \tag{11}$$

In these expressions:
v is the polymer concentration by volume in the swollen gel (i.e., the reciprocal of the equilibrium volume swelling ratio: Q^{-1});
χ is the polymer solvent interaction parameter, which is a dimensionless quantity characterizing a polymer solvent system. χ may generally depend on temperature, pressure, concentration, but only a little on the functionality f;
$h^{2/3}$ is the memory term, already discussed;
V_{01} is the molar volume of the solvent;
v is the number of elastically effective network chains, in moles per unit volume of dry network.

If the network — obtained by endlinking — is ideal, all precursor chains have been converted into elastically effective network chains, and v can be expressed in terms of the molecular weight M of the elastic chains by:

$$v = (M\ \tilde{V}_{02})^{-1}$$

where $\tilde{V}_{02}$ is the specific volume of the dry network.

The swelling equation has been written in a form which summarizes the various theoretical expressions[c] by using the factors A and B. Flory and Wall[38] came to a result which is formally obtained if A is set equal to 1 and B to $2/f$ in Eq. (10). The corresponding values according to the theory of Hermans[39] are: $A = 1$ and $B = 1$. Duiser and Stavermann[40] and later Graessley[41, 42] derived expressions by treating small parts of networks consisting of limited numbers of chains and of nodules. Their expression results in the following values: $A = 1 - 2/f$; $B = 1/2$. In a recent contribution, Flory[43] confirms the result of James and Guth[44, 45] for $B = 0$, and that of Graessley[41, 42] with $A = 1 - 2/f$, for a "phantom" network, the elastic chains of which do not exhibit any excluded volume and are allowed to move freely through one another. For a tetrafunctional phantom network Flory gets thus $A = 1/2$, a value which had been derived independently by Eichinger[46] and by Edwards[47]. For real networks, in which excluded volume has to be considered and in which the movements of the junctions about their mean position are restricted by interactions with neighboring chains, and possibly by entanglements, Flory[43] argues that, except for high swelling ratios, B should not vanish.

It should be emphasized that if B is supposed to be equal to zero (or very small), A should contain the functionality of the crosslinks, since it is obvious that f plays a role in the swelling equilibrium.

From Eq. (11), and taking into account the expression $v = (M \tilde{V}_{02})^{-1}$, valid for ideal networks, it can be seen that the analytical relationship between the equilibrium swelling degree Q and the molecular weight M of the elastic chains is rather complicated. Several additional points have to be considered:

χ is a function of the segment concentration v[48].

Series expansion of $\ln(1 - v)$ and of $v^{1/3}$ have to be carried out to higher terms when the segment concentration is not very small, *i.e.*, when Q is not very high.

Even if the experimental points fit a line $Q = K M^{\alpha}$ it cannot be expected that it passes through the origin.

5.2. Discussion of Experimental Results

We may distinguish between networks exhibiting high swelling ratios (Figs. 5a and b), and networks which do not swell much (Fig. 6).

5.2.1. Networks with High Swelling Ratios

Polystyrene model networks swollen in benzene are a typical example of networks exhibiting high swelling ratios. Experimentally it was found that Q varies linearly with $M^{3/5}$ for networks covering a wide range of elastic chain lengths[29, 34]. However, Rietsch and Froelich[36, 49] claim that they have observed a linear relationship between Q and M. Their gels were obtained by endlinking processes, at rather low

---

[c]   It is worth mentioning that some authors – such as Tobolsky[33] – refer to the product $A \langle r^2 \rangle / \langle r_{05}^2 \rangle$ as the front factor, whereas others call front factor the sole constant A.

concentrations. It should be added that the values of Q which they obtained are not really contradictory with a proportionality law between Q and $M^{3/5}$ (Fig. 5a).

To account for the relationship experimentally observed we may introduce the expression for ν into Eq. (11); developing the logarithm to the second order one gets:

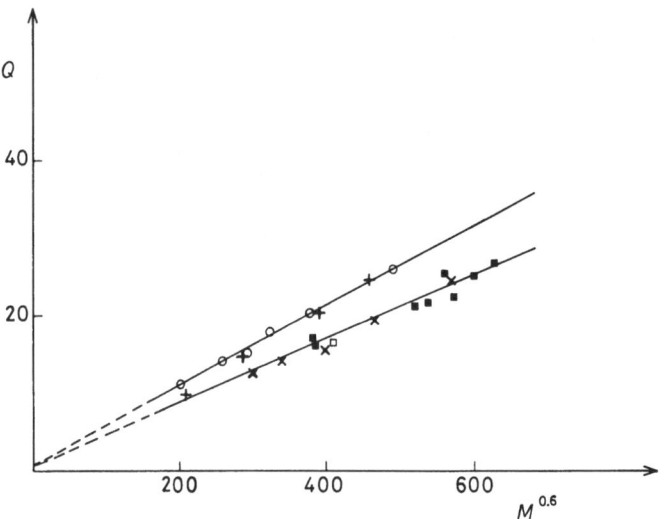

Fig. 5 (a). Equilibrium swelling degree Q of PS networks in benzene vs. $M^{3/5}$ (M in g · mol^{-1})
o: $f = 3$ $v_c = 0.15$[14]
+: (DVB)/(LE) = 3 $v_c = 0.07$[36, 38]
■: (DVB)/(LE) = 4 $v_c = 0.11$[27]
x: (DVB)/(LE) = 3 $v_c = 0.10$[14]

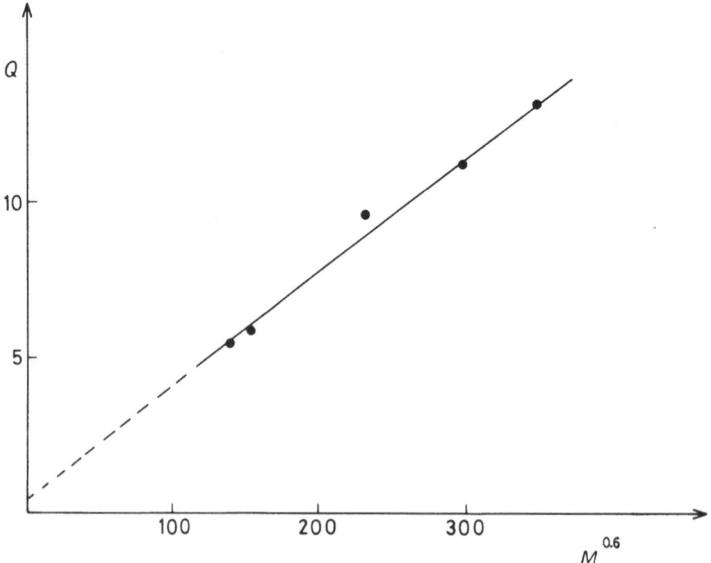

Fig. 5 (b). Equilibrium swelling degrees Q of polydimethylsiloxane (PDMS) networks in heptane vs $M^{3/5}$ (M in g · mol^{-1})

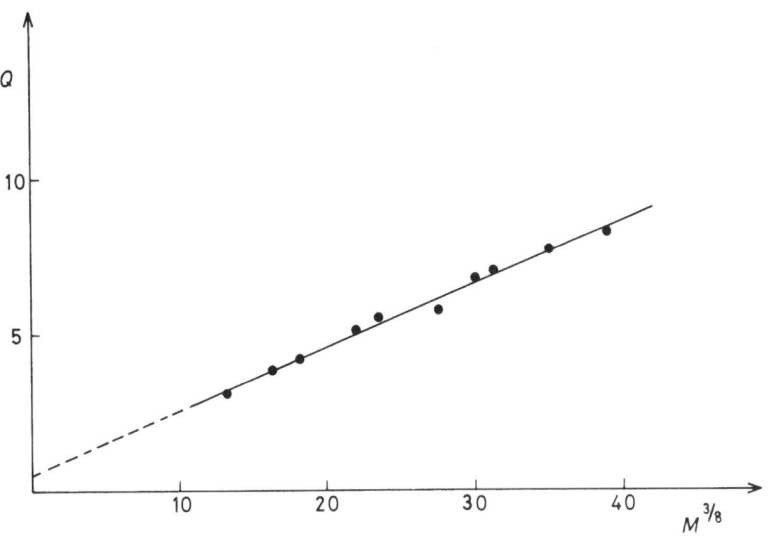

Fig. 6. Equilibrium swelling degrees Q of PDMS networks in toluene vs $M^{3/8}$ (M in g · mol^{-1})

$$\frac{1}{M\, \tilde{V}_{02}} = \frac{(1/2 - \chi)v^2}{V_{01}(A\, v^{1/3}h^{2/3} - B\, v)} \tag{12}$$

Since $v = Q^{-1}$

$$M\, \tilde{V}_{02} = V_{01}\, \frac{(A\, h^{2/3} - B\, Q^{-2/3})}{1/2 - \chi}\, Q^{5/3}$$

or

$$Q = \left[\frac{\tilde{V}_{02}}{V_{01}}\, \frac{1/2 - \chi}{A\, h^{2/3} - B\, Q^{-2/3}}\right]^{3/5} M^{3/5} \tag{13}$$

The experimentally observed proportionality between Q and $M^{3/5}$ requires that the difference $(A\, h^{2/3} - B\, Q^{-2/3})$ varies only very little with the molecular weight of the elastic chains or the swelling degree, respectively. If the segment concentration of the nascent network is taken as the swollen reference state, h can be considered as a constant. This implies that either B is zero, or $B\, Q^{-2/3}$ is very small compared with $A\, h^{2/3}$, provided χ is a constant. If h is not a constant — this will be discussed in the next section — one has to assume that the molecular weight dependence of $A\, h^{2/3}$ and of $B\, Q^{-2/3}$ are compensating, in order for the difference to remain independent of the molecular weight.

More elaborate treatments of the swelling data were attempted[50] without great success, since besides the two constants A and B a third parameter, χ, is not known precisely. One may assume that χ remains nearly the same, whether the polymer is linear, or branched, or even crosslinked, since it merely characterizes the interactions between remote segments in the swelling solvent considered. This has been confirmed recently by Candau, Strazielle and Benoit[51]: according to their results the χ_* value for star-shaped polystyrene in benzene is only slightly higher than the χ_{lin} value for linear homologues. Moreover, the difference between χ_* and χ_{lin} was shown to decrease as the segment concentration increases. It therefore seems justified to assume that the χ value for a network is equal to that of a branched (or even of a linear) polymer, measured at the same segment concentration v, provided v is high enough.

Measurement of χ at rather high segment concentration may be performed either by light scattering[52] or by ultracentrifugal techniques[53]. Swelling measurements combined with uniaxial compression measurements also allow in principle the determination of χ, but high experimental accuracy is required. Froelich[36] applied this method to polystyrene model networks and confirmed that there is little or no difference between $\chi_{network}$ and χ_{lin} for polystyrene swollen in benzene. Moreover χ is found to be independent of the swelling degree (i.e., of the molecular weight of the elastic chains). However when swollen in poor solvents (cyclohexane) χ depends strongly upon the swelling degree. Its variation with the volume segment concentration $v(= Q^{-1})$ and with temperature is described by a relationship similar to that of Scholte[54].

All these considerations explain why attempts quantitatively to interpret the observed values of the slope of Q vs $M^{3/5}$ lines have failed. If the slope is calculated using χ values found in the literature, assuming that $B = 0$ or $B = 1/2$ and that A is related to the functionality according to Graessley[41, 42]:

$$A = 1 - 2/f$$

the best fit with experiment is obtained for $h^{2/3} = 0.5$ and $B = 0$. But this result also involves a hypothesis on the value of f which was set equal to 4[35].

5.2.2. Networks with Low Swelling Degrees

When the equilibrium volume swelling degree is low — as it is for PDMS networks obtained at very high polymer concentration (Fig. 6) — it is found experimentally that Q is a linear function of $M^{3/8}$. To account for this result one has to develop the logarithmic term in Eq. (11) to the third order.

This yields:

$$\frac{1}{M\, \tilde{V}_{02}} = \frac{(1/2 - \chi)v^2 + v^3/3}{V_{01}[A\, h^{2/3} v^{1/3} - B\, v]} \tag{14}$$

or

$$M\, \tilde{V}_{02} = \frac{V_{01}(A\, h^{2/3} - B\, v^{2/3})}{(1/2 - \chi)1/v + 1/3}\, v^{-8/3}$$

resulting in:

$$Q = \left[\frac{\tilde{V}_{02}}{V_{01}} \cdot \frac{\frac{1}{2} - \chi \ Q + \frac{1}{3}}{A \, h^{2/3} - B \, Q^{-2/3}} \right]^{3/8} \cdot M^{3/8} \tag{15}$$

It has thus to be assumed that the difference $(A \, h^{2/3} - B \, Q^{-2/3})$ is independent of the molecular weight and that $(1/2 - \chi) \, Q \ll 1/3$. The first of these conditions has already been discussed in the preceeding paragraph. The second condition requires χ to be close to $1/2$. This assumption is plausible, since it was shown[55, 56] that at high concentrations, in any solvent, the expansion of the chains decreases, and consequently χ tends to $1/2$, as it does at the Θ-point.

It is thus possible to account for the type of variation of Q with the molecular weight of the elastic chains, but a quantitative check is doubtful in most cases, because of the number of parameters which have to be introduced.

6. Unidirectional Deformation of Model Networks

6.1. Some Theoretical Considerations

A mechanical deformation induces a change of the free energy of the network chains, which is given by the well-known expression:

$$\Delta G_{\text{el}} = -\nu R T \left[B \ln(\lambda_x \lambda_y \lambda_z) - \frac{A}{2} (\lambda_x^2 + \lambda_y^2 + \lambda_z^2 - 3) \right] \tag{16}$$

in which λ_x, λ_y, and λ_z are the average deformation ratios of the ν elastically effective network chains of the network[1, 57]. These parameters are related to the macroscopic deformation ratios of the gel, $\Lambda_x, \Lambda_y, \Lambda_z$ by

$$\lambda_i = \left(Q \frac{b_d}{b_0} \right)^{1/3} \cdot \Lambda_i ; \quad i = x, y, z$$

Here b_d and b_0 are the average volumes occupied by an elastic chain in the dry undeformed state and in the swollen reference state, respectively. The ratio b_d/b_0 should therefore be identical with the term h, previously discussed.

The variation of the elastic free energy between the elastically deformed state and the initial undeformed state is the work accomplished by the force F applied to the sample. For a uniaxial deformation along the x axis yielding a macroscopic deformation ratio Λ_x:

$$\Delta G_{\text{el}} - \Delta G_{\text{el}, i} = \frac{\nu A R T}{2} \, Q_i^{2/3} h^{2/3} \left[\Lambda_x^2 + 2 \frac{Q}{Q_i} \Lambda_x^{-1} - 3 \right]$$

This expression takes into account that:

$$\Lambda_y^2 = \Lambda_z^2 = \frac{V}{V_i} \Lambda_x^{-1} = \frac{Q}{Q_i} \Lambda_x^{-1}$$

where Q and Q_i are the actual swelling ratios (in the uniaxially deformed state) and the initial one; V and V_i are the corresponding sample volumes. If we call F the applied force, V_d the volume of the dry sample, L_i its initial length and Λ the deformation ratio along the x axis ($\Lambda = L/L_i$) we may write:

$$F = \nu A RT \frac{V_d}{L_i} Q_i^{2/3} h^{2/3} \left[\Lambda - \frac{Q}{Q_i} \Lambda^{-2} \right] \tag{17}$$

We may now define a modulus of deformation per unit area of unstrained unswollen gel, to allow for comparison between homologous networks exhibiting various initial swelling degrees:

$$G^* = \frac{F}{\frac{V_d}{L_i} Q_i^{2/3} \left(\Lambda - \frac{Q}{Q_i} \Lambda^{-2} \right)} = A RT h^{2/3} \nu \tag{18}$$

Q/Q_i characterizes the deswelling effect imposed on the network by the applied force. If the gel does not deswell this ratio is equal to unity. The deswelling, for large samples, is a very slow process. Therefore it is often difficult to decide whether the new swelling equilibrium has been attained or not. But it is possible[58] to make use of an extrapolation procedure. The deformation is measured as a function of time, starting from times which are not too short, to avoid frictional effects. The extrapolation to zero time yields a Λ value which is not influenced by any diffusion process (yielding deswelling by compression).

6.2. Comparison with Experimental Data

6.2.1. Variation of G^* with M

Since the reduced modulus G^* refers to the unswollen, unstrained, isotropic network, the question arises, first, whether it should be the same for a given network, regardless of the swelling solvent. It was found that for PDMS networks as well as for polystyrene networks the value of G^* is not the same for a given gel whatever the swelling solvent may be[14, 22]. This result, which implies that the swelling solvent influences the mechanical response of the network, is still somewhat controversial[59], however.

The modulus $G*$ defined above is proportional to the number of elastically effective network chains ν, and to the 2/3 power of the term h. If the networks investigated are sufficiently close to ideality, the number ν of network chains is equal to the number of precursor chains and ν may be replaced by $(M\tilde{V}_{02})^{-1}$, yielding:

$$G* = \frac{A\,RT}{\tilde{V}_{02}}\,\frac{h^{2/3}}{M} \tag{19}$$

It is therefore of interest to study the variation of $G*$ with the molecular weight M of the elastic chains, for a homologous series of model networks swollen in one given solvent, and submitted to uniaxial deformation. If the memory term $h^{2/3}$ is a constant for a given gel, $G*$ should be a linear function of the reciprocal of the molecular weight of the elastic chains. This result was found by Froelich and Rietsch[59]; however, Herz and Belkebir-Mrani[14, 22] established that $G*$ is proportional to $M^{-\alpha}$, where α is of the order of 1.5 for thermodynamically "good" swelling solvents, and of the order of 4/3 for Θ-solvents. These findings are rather well established, and hold for polystyrene as well as for polydimethylsiloxane model networks. They are obviously contradictory with the assumption – discussed earlier – that h should be an intrinsic property of the network, related solely to the conditions of its preparation. To be more precise, one can state that the relationship between $G*$ and M for

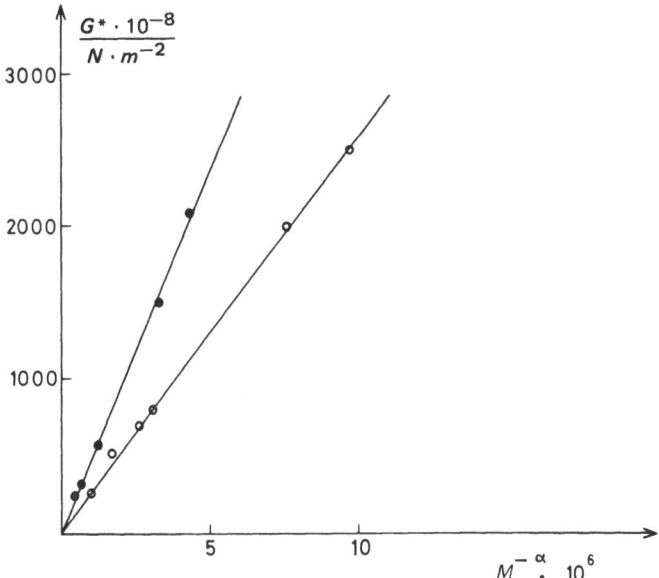

Fig. 7. Modulus $G*$ of PDMS networks ($f = 4$) vs molecular weight (M in g · mol^{-1}).
●: heptane, $\alpha = 1.5$
○: toluene, $\alpha = 1.4$

a series of networks does not characterize the solvent in which the preparation of the networks was carried out[29], but merely the swelling solvent: PDMS networks synthesized in toluene (a rather "bad" solvent) and swollen in heptane (a well-solvating solvent for the elastic chains) yield an exponent of −1.5, typical for good solvents. Conversely PDMS networks prepared in heptane and swollen in toluene exhibit a $G*$ vs M relationship characteristic for Θ-solvents (above Θ-temperature) with an exponent of the order of −1.4. (Fig. 7).

We shall now try to account for these results, assuming, as we did previously, that within a series of networks the functionality of the crosslinks remains constant an assumption which seems to rest on solid grounds.

Recalling the definition of the memory term $h^{2/3}$, as given by Dusek and Prins[3]:

$$h^{2/3} = \frac{\langle r_d^2 \rangle}{\langle r_{os}^2 \rangle}$$

we shall try to evaluate roughly the variation of both terms $\langle r_d^2 \rangle$ and $\langle r_{os}^2 \rangle$ with M.

$\langle r_d^2 \rangle$: the mean square end-to-end distance of the elastic chains in the unswollen swollen network can be deduced from Eq. (9):

$$\langle r_d^2 \rangle = x^2 d^2 = C^2 x^2 \left(\frac{f}{2 \rho_2 \mathcal{N}} \right)^{2/3} M^{2/3} \tag{20}$$

and it is thus found that $\langle r_d^2 \rangle$ should be proportional to $M^{2/3}$ if neither the geometric functionality — which determines the value of C — nor the connectivity factor x^2 — characterizing the proportion of chains linking crosslinks which are *not* first neighbors — varies within a series of homologous model networks.

$\langle r_{os}^2 \rangle$: the mean square end-to-end distance of the corresponding free chains should be taken in the swelling solvent, and not in the solvent in which the network synthesis was carried out. It can be roughly estimated from the viscosity law of the corresponding linear polymer in the same solvent:

$$\langle r_0^2 \rangle = \left(\frac{K}{\Phi} \right)^{2/3} M^{2(1+a)/3} \tag{21}$$

Here K and a are the parameters of the Mark-Houwink-Sakurada viscosity law:

$$[\eta] = K M^a$$

and Φ is Flory's universal parameter.

It follows that, under the assumption made, $h^{2/3}$ should be proportional to $M^{-2a/3}$. Consequently $G*$ can be expressed as:

$$G* = A\, RT\, C^2 x^2 \left(\frac{f}{2 \rho_2 \mathcal{N}} \frac{\Phi}{K} \right)^{2/3} M^{-(1+2a/3)} \tag{22}$$

If the value of a is of the order of 0.6 (Θ-solvents, above Θ-temperature) G^* should be proportional to $M^{-1.4}$, as was found experimentally; in very good swelling solvents, where the exponent a is of the order of $3/4$, G^* should be proportional to $M^{-3/2}$, in good agreement with experimental findings (Fig. 7).

The above treatment requires some discussion on several points:

We already mentioned that this expression originates from a definition of h which is incompatible with that used in the first section of this review. Here the swelling solvent is considered instead of the solvent in which crosslinking occurred. This point is obviously not settled and deserves more experimental work.

The intercrosslink distance d is found experimentally to be linear in $M^{0.4}$ instead of $M^{1/3}$ for unswollen networks, as mentioned earlier.

It should be emphasized also that the viscosity relation applies to infinitely dilute solution. In a network swollen to equilibrium the segment concentration is far from being infinitely small.

The above treatment is therefore a very rough approximation, and has to be considered as such. Nevertheless the fit between the observed and the calculated exponent α of the relations:

$$G^* = K' M^{-\alpha}$$

is rather good, and quantitative comparisons may even be undertaken, concerning the slopes of the lines, as will be shown below.

6.2.2. Influence of the Crosslink Functionality

Let us compare the behavior of networks exhibiting different crosslink functionalities. Qualitatively it is known that an increase of f results in a small decrease of the swelling ratio Q – at constant molecular weight of the elastic chains[34] –, and in a rather sharp increase of the reduced modulus G^*[22].

For each homologous series of networks, swollen in a given solvent, G^* is plotted versus $M^{-\alpha}$, yielding a straight line (Fig. 8a, b). The higher the average crosslink functionality characterizing the series, the steeper the slope of the lines. Using the trifunctional networks[d] as reference, one may thus try to calculate the functionalities f of the other series of networks from the obtained slopes, which are increasing as $(A \cdot f^{2/3})$, according to Eq. (22). If A is taken equal to $(1 - 2/f)$, according to Graessley[41, 42], one can write:

$$F(f) = \frac{\text{slope}_f}{\text{slope}_3} = 3^{1/3} \cdot f^{-1/3} (f - 2) \tag{23}$$

Comparison is thus rendered possible between networks with elastic chains of the same average length, but exhibiting different crosslink functionalities. From the ratio

---

[d]   The trifunctional polystyrene networks are the only that are made by stoichiometric deactivation using tris(allyloxy)-s-triazine. They are the only polystyrene networks the functionality of which is known with some accuracy.

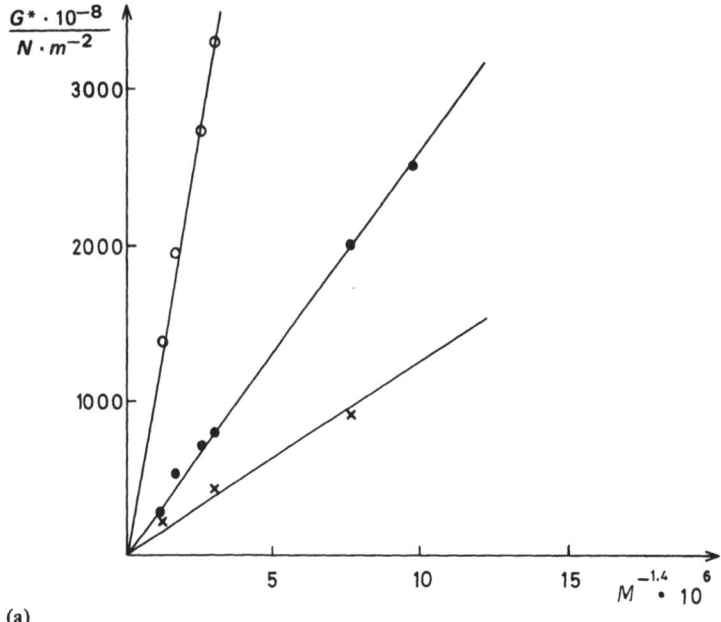

(a)

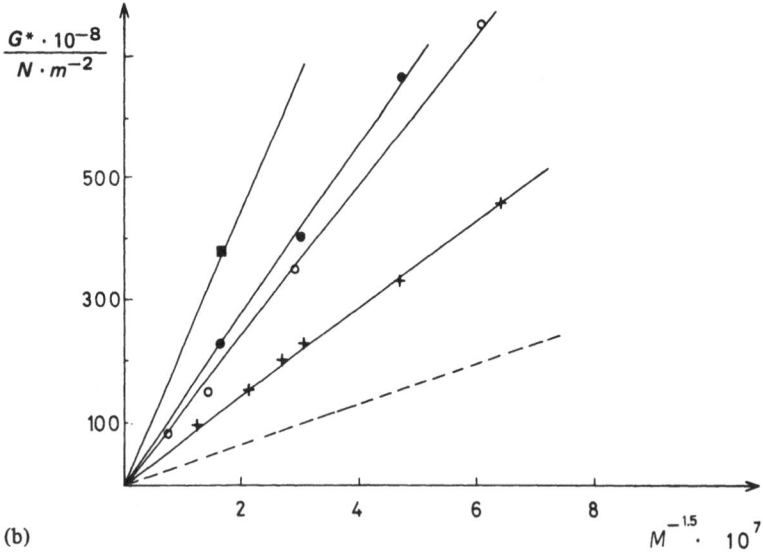

(b)

Fig. 8 (a). Modulus G^* of PDMS networks with 3, 4, and 6 functional crosslinks vs molecular weight $M^{-1.4}$. (M in g · mol^{-1}).

Solvent: toluene

x: $f = 3$; •: $f = 4$; o: $f = 6$

Fig. 8 (b). Modulus G^* of PS networks prepared with different [DVB]/[LE] ratios vs molecular weight $M^{-1.5}$; solvent: benzene.

[DVB]/[LE]: +: 3; o: 5; •: 6; ■: 10

dotted line: $f = 3$

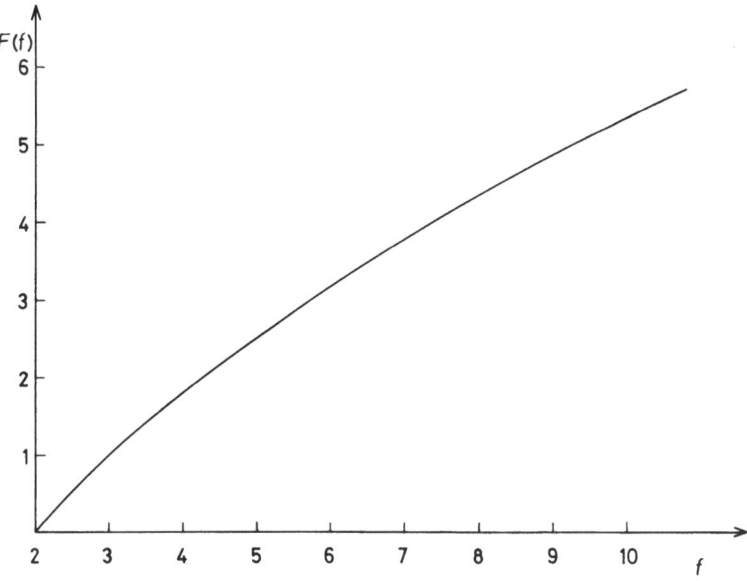

Fig. 9. $F(f)$ vs f according to Eq. (23)

Table 2. Comparison of PS networks prepared with various amounts of DVB with three-functional PS networks containing elastic chains of comparable length. $f_{calc.}$ has been determined from the $F(f)$ curve in Fig. 9

DVB/LE	$\dfrac{M}{g \cdot mol^{-1}}$	$\dfrac{G^* \cdot 10^{-8}}{N \cdot m^{-2}}$	$G^*_{DVB}/G^*_{f=3}$	$f_{calc.}$	$\dfrac{G^*_{f=3} \cdot 10^{-8}}{N \cdot m^{-2}}$	
3	13500	457	1.94	4.1	235	$(M \approx 13000)$
5	14000	755	3.21	6		
3	22000	230	2	4.2		
5	22500	350	3.04	5.8	115	$(M \approx 20000)$
6	22000	406	3.54	6.6		
6	33500	229	3.52	6.6	65	$(M \approx 30000)$
10	28000	378	5.82	~11		

of the slopes of the G^* vs $M^{-\alpha}$ lines, the values of the functionality f can be calculated, assuming the reference networks to be really trifunctional (Fig. 9). From the values shown in Table 2 it can be seen that up to crosslink functionalities of 5 or 6 the results are self-consistent.

This means that the connectivity factor x^2 – which was not taken into account – should be close to unity: the vast majority of the elastic chains connect first neighbor crosslinks, as long as the average functionality does not exceed 5 or 6. For PDMS networks the fit is only good up to $f = 4$. For polystyrene networks the limiting functionality is attained for a proportion of DVB per living end of the order of 5. (Table 3).

Table 3. Comparison of PDMS networks with known functionalities of crosslinks

$f_{exp.}$	$\dfrac{M}{g \cdot mol^{-1}}$	$\dfrac{G^* \cdot 10^{-8}}{N \cdot m^{-2}}$	$G^*/G_{f=3}$	$f_{calc.}$
3		900		
	4,500			
4		2,025	2.25	4.6
3		435		
4	8,700	835	1.9	4.1
6		3,305	7.6	very high
3		235		
4	17,100	270	1.20	3.2
6		1,385	5.9	very high

$f_{calc.}$ has been determined in the same way as for PS networks. Obviously when $f = 6$, the calculated value of $f_{calc.}$ is not compatible with the known value: x^2 has to be greater than unity and moreover the entanglement contribution has become very important.

For networks in which the average crosslink functionality exceeds the above-mentioned values, the ratios of the slopes lead to functionalities which are obviously too high. The assumption according to which the connectivity factor x^2 is equal to unity cannot hold any more. This means that a fraction of the network chains connect nodules which are *not* first neighbors. The probability for permanent entanglements to occur is increased. No quantitative treatment of the experimental data is possible in that case, because of the number of parameters to consider, none of them being accessible by independent experiments.

It is worth mentioning that these findings are consistent with very recent results obtained by inelastic light scattering experiments carried out on the same polystyrene and polydimethylsiloxane networks as used in uniaxial deformation studies[60, 61]. The cooperative diffusion constants of these networks swollen to equilibrium were investigated by Raleigh light scattering line width measurements, as a function of the average length of the elastically effective network chains and of the crosslink

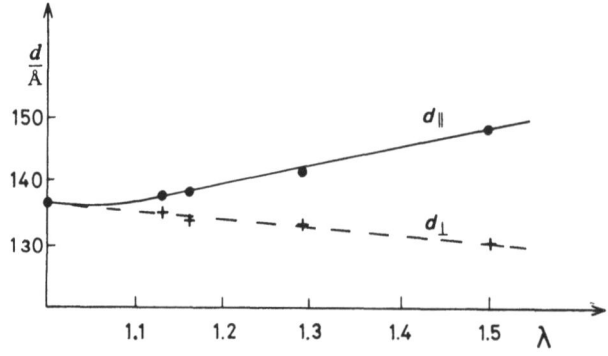

Fig. 10. Variation of the correlation distances $d_{\parallel}$ and $d_{\perp}$ vs elongation ratio Λ

functionality. The results obtained lead to the conclusion that for networks with high f values chain interpenetration becomes rather important: a contribution from physical entanglements clearly appears. On the contrary in networks in which f is of the order of 4 the contribution of entanglements to the network behavior becomes negligible. For trifunctional networks this technique is capable of detecting the influence of defects, such as pendant chains and chain coupling. A slightly increased proportion of such defects was found as expected — for networks prepared by stoichiometric reaction with trifunctional reagents. Light scattering spectroscopy of networks may be a very efficient tool for investigating the structure of networks, and it has confirmed some of the conclusions obtained from swelling and uniaxial deformation measurements[60, 61].

6.3. Uniaxial Deformation of Labelled Networks

Neutron scattering experiments were also carried out on some model networks with labelled crosslinks, in the unswollen state[13, 26]. To achieve such measurements the dry polystyrene networks are heated above their glass transition temperature, submitted to uniaxial deformation, quenched under stress and studied in the neutron scattering apparatus.

As can be expected, the angular scattering intensity distribution has become anisotropic. From the data obtained, two values of the average correlation distance between neighboring nodules can be calculated, parallel and perpendicular to the deformation direction respectively. Although the accuracy of the data is not outstanding, it can be seen that $d_{\parallel}$ increases with the macroscopic extension ratio, Λ_x, whereas $d_{\perp}$ decreases only slightly (Fig. 10). It is not possible to ascertain experimentally whether $d_{\perp}$ is proportional to $\Lambda_x^{-1/2}$ as it should be if the deformation proceeded at constant volume i. e., if the network could be considered incompressible.

Similar results[26], yielding even better accuracy have been obtained with swollen networks, submitted to an extension Λ. Further experimental work is necessary to settle this point. Nevertheless the increase of $d_{\parallel}$ with Λ_x is a further argument in favor of the "spring-suspended beads" model for the networks obtained by endlinking processes, as it is a direct proof of the proportionality between the macroscopic deformation and the deformation of the individual structural elements in the network, when it is submitted to a stress.

7. Concluding Remarks

This review was devoted to tridimensional polymeric networks of well-defined structure, called for that reason model networks. These networks are synthesized by endlinking processes, whereby well-characterized linear precursor chains become elastically effective network chains. The model network can be considered close to ideality, as the number of defects such as dangling chains, loops, and double connections is generally small.

The major advantage of the endlinking processes is that the porosity of the networks can be chosen at will: precursor chains covering a wide range of molecular weights – and exhibiting generally a sharp molecular weight distribution – can be used for the synthesis of the model networks. Moreover labelled gels can be made and have been studied, thus giving direct access to the average distance between the structural elements, a goal which could not be reached with classical networks.

It was beyond the scope of this review to describe the influence of defects such as dangling chains on the properties of networks. It should be mentioned nevertheless that the methods described above also allow the synthesis of networks with a controlled proportion of defects. By using a well-defined mixture of mono- and difunctional initiator species one may get precursor chains with either one or two living ends. If this precursor solution is treated with a known amount of divinylbenzene, the reaction medium gels, but each of the monofunctional precursor chains becomes a dangling chain, attached to the network. The proportion of monofunctional initiator used defines the proportion of dangling chains. Such networks have been studied by Bastide and Picot[62]. It was thus confirmed that the modulus G^* is much more sensitive to defects of the networks than the equilibrium swelling degree, in a given solvent. Whereas Q hardly changes upon increasing the number of dangling chains, G^* decreases as expected. Similar effects had been observed on PDMS networks by Belkebir-Mrani[22].

Along the same lines, it should be mentioned that model networks can be synthesized with linear polymer chains trapped in the tridimensional matrix. If the chains are chosen long enough they cannot diffuse out of the network, even when it is swollen in good solvents. With respect to similar gels containing no trapped chains, the swelling degree is unchanged and the modulus G^* is reduced slightly. The relaxation behavior of the trapped chains has been studied by means of stress strain measurements on dry networks and on networks swollen in "bad" solvents[63]. It would be of interest to establish whether the predictions of De Gennes[64] concerning the "reptation" of free polymer chains in a tridimensional matrix can be checked by experiments adequately chosen.

Homologous series of networks, prepared under the same experimental conditions, and differing only by the length of their elastic chains have been made available, and were used to investigate their structure, their swelling behavior, and their mechanical deformation properties. It was assumed that within a homologous series of networks the crosslink functionality remains constant, an assumption which is supported by results on star-polymers[11].

The most important result obtained on the structure of model networks is the existence of a rather well-defined correlation distance between first neighbor crosslinks. The affine character of deformations induced by swelling processes is clearly shown by the fact that, for a given gel, the correlation distance d increases as the cubic root of the volume swelling degree Q up to values of Q of the order of 10. A model network can thus be schematized by an ensemble of spring-suspended beads: the elastic chains act as entropy springs, and they connect beads of constant – as yet unknown – functionality.

If the functionality of the crosslinks is low, the probability for elastic chains to connect first neighbor crosslinks is close to unity. Consequently there is little inter-penetration of the elastic chains and few permanent entanglements are expected. However this does not mean that *all* first neighbor crosslinks in the gel are connected by elastic chains.

If the functionality of the branch points is of the order of 5 (or higher) the probability for elastic chains to connect branch points which are *not* first neighbors cannot be neglected any more. The mean square end-to-end distance of the elastic chains $\langle r^2 \rangle$ is then higher than the square of the average distance d between neighbor crosslinks. The ratio $x^2 = \langle r^2 \rangle / d^2$ characterizes the degree of interpenetration of the network chains. The probability of occurrence of permanent entanglements in the gel should increase when x^2 increases: however this statement is as yet merely qualitative.

For networks belonging to a homologous series the average distance between neighbor crosslinks increases when the length of the elastic chains in the network increases. The increase is more rapid for swollen gels than for dry networks owing to the fact that swelling is itself a function of the size of the elastic chains. But a satisfactory agreement between theoretical expectations and experimental results was found. The swelling behavior of the networks can be accounted for by the classical expressions, assuming that the reference state characterizes the swelling ratio Q_c of the nascent network: The conformational changes undergone by the chains upon crosslinking are small and the ratio of the mean square end-to-end distance of the elastic chains to the mean square end-to-end distance of the free chains should be close to unity. The memory term thus characterizes the nascent network, *i.e.*, the degree of swelling it had when it was formed.

The same value of $h^{2/3}$ should intervene in the expression of the modulus as defined from uniaxial compression measurements. But it was clearly established that the response of a swollen gel to a mechanical stress depends upon its actual swelling solvent and upon its actual swelling ratio, and not upon the conditions of network formation. Here a difficulty arises. Account can be given of the experimental values of the modulus if the term h solely refers to the swelling solvent, whereas for swel-ling ratios the value of h was referring to the state of swelling of the nascent network. This discrepancy is not easy to explain and more experimental work is needed to settle the problem.

This review does not include a discussion of the actual porosity of the networks. However, it should be stated that investigations were carried out to characterize the equilibrium partition coefficients in ternary systems: solvent/polymer/model-net-work. A rather important deswelling effect is observed when a network is immersed in a polymer solution, with respect to the swelling degree the network exhibited in the pure solvent[65]. Therefore no precise information on the pore size distribution can be obtained from the partition curves. Gel permeation chromatography experi-ments can also be carried out using model networks, in spite of the fact that they do not exhibit any macroporosity — no syneresis occurred upon crosslinking —. How-ever, the permeation is a slow process and the static and dynamic partition coefficients differ widely[66].

Finally, it is also worth mentioning that measurements of glass transition temperatures were carried out on model networks. It was shown that T_g increases linearly with the reciprocal of the molecular weight of the elastic chains, for a series of homologous networks. The slope of the lines is the steeper, the higher the functionality of the crosslinks[67].

Model networks can be considered as a new class of tridimensional polymeric material, with elastic chains of known and rather constant length, and branch points of almost constant functionality, with few defects, and of very satisfactory homogeneity. By combination of structural determinations and of swelling and deformation measurements it was shown that the existing theories of rubber elasticity are applicable, even though some ambiguity still exists as to what concerns the value and the meaning of the memory term.

Even though much work has still to be done to get a more precise insight into the behavior of crosslinked polymers, one can state that decisive progress has been accomplished along this line, and can be further expected, by the use of model networks.

8. References

[1] Flory, P. J.: Principles of polymer chemistry. Ithaca (NY): Cornell University Press 1953
[2] Seidl, J., Malinsky, J., Dusek, K., Heitz, W.: Adv. Polym. Sci. 5, 113 (1967)
[3] Dusek, K., Prins, W.: Adv. Polym. Sci. 6, 1 (1969)
[4] Dusek, K.: J. Polym. Sci. C-16, 1289 (1967)
[5] Chen, R. Y. S., Yu, C. U., Mark, J. E.: Macromol. 6, 746 (1973)
[6] Walsh, D. J., Allen, G., Ballard, G.: Polymer 15, 366 (1974)
[7] Szwarc, M.: Carbanions, living polymers and electron transfer processes. New York: Interscience 1968
[8] Weiss, P., Hild, G., Herz, J., Rempp, P.: Makromol. Chem. 135, 249 (1970)
[9] Beinert, G., Belkebir-Mrani, A., Herz, J., Hild, G., Rempp, P.: Faraday Discus. Chem. Soc. 57, 27 (1974)
[10] Dudek, B., Plominka, B.: XXIII Intern. Symposium on Macromolecules (IUPAC) Madrid (1974), p. 2103
[11] Worsfold, D. J., Zilliox, J. G., Rempp, P.: Canad. J. Chem. 47, 3379 (1969)
[12] Belkebir-Mrani, A., Beinert, G., Herz, J., Mathis, A.: Europ. Polym. J. 12, 243 (1975)
[13] Benoit, H., Decker, D., et al.: J. Polym. Sci. (Phys. Ed.) 14, 2119 (1976)
[14] Belkebir-Mrani, A., Herz, J., Rempp, P.: Makromol. Chem. 178, 485 (1977)
[15] Herz, J., Hert, M., Strazielle, C.: Makromol. Chem. 160, 213 (1972)
[16] Strazielle, C., Herz, J.: Europ. Polym. J. 13, 223 (1977)
[17] Pinazzi, C., Esnault, J., Lescuyer, G., Villette, J.-P., Pleurdeau, A.: Makromol. Chem. 175, 705 (1974)
[18] Mark, J. E., Sullivan, J. L.: to be published
[19] Uraneck, C. A., Hsie, H. L., Buck, O. G.: J. Polym. Sci. 46, 535 (1960)
[20] Kraus, G., Moczygemba, G. A.: J. Polym. Sci. A-2, 277 (1964)
[21] Herz, J., Belkebir-Mrani, A., Rempp, P.: Europ. Polym. J. 9, 1165 (1973)
[22] Belkebir-Mrani, A., Beinert, G., Herz, J., Rempp, P.: Europ. Polym. J. 13, 277 (1977)
[23] Hopkins, W., Peters, R. H., Stepto, R. F. T.: Polymer 15, 315 (1974)
[24] Allen, G., Egerton, P. L., Walsh, D. J.: Polymer 17, 65 (1976)
[25] Graessley, W. W.: Adv. Polym. Sci. 16, 1 (1974)
[26] Duplessix, R.: Thesis, Strasbourg (1975)

27) Lutz, P., Picot, C., Hild, G., Rempp, P.: Brit. Polym. J.; to be published
28) Strazielle, C., Benoit, H.: Macromol. 8, 203 (1975)
29) Rempp, P., Herz, J., Hild, G., Picot, C.: Pure and Appld. Chem. 43, 77 (1975)
30) Ziabicki, A.: Colloid and Polym. Sci. 252, 49 (1974)
31) Cotton, J. P., Decker, D., et al.: Macromol. 7, 863 (1974)
32a) Levy, S.: Thesis, Strasbourg (1964)
  b) Volkenstein, M. V.: Configurational statistics of polymeric chains. New York: Interscience 1963
33) Green, M. S., Tobolsky, A. V.: J. Chem. Phys. 14, 80 (1946)
34) Weiss, P., Herz, J., Rempp, P.: Makromol. Chem. 141, 145 (1971)
35) Froelich, D., Crawford, D., Rozek, R., Prins, W.: Macromol. 5(1), 100 (1972)
36) Rietsch, F., Froelich, D.: Europ. Polym. J., to be published
37) Haeringer, A., Hild, G., Rempp, P., Benoit, H.: C. R. Acad. Sci. C-276, 1711 (1973)
38) Flory, P. J., Wall, F. T.: J. Chem. Phys. 19, 1435 (1951)
39) Hermans, J. J.: Trans. Faraday. Soc. 43, 591 (1947)
40) Duiser, J. A., Staverman, J. A.: Physics of non-crystalline solids, Prins, J. A., ed. Amsterdam: North Holland 1965, p. 276
41) Graessley, W. W.: Macromol. 8, 186 (1975)
42) Graessley, W. W.: Macromol. 8, 865 (1975)
43) Flory, P. J.: Proc. Royal Soc. London A-351, 351 (1976)
44) James, H. M., Guth, E.: J. Chem. Phys. 11, 455, 472 (1943)
45) James, H. M., Guth, E.: J. Chem. Phys. 21, 1039 (1953)
46) Eichinger, B.: Macromol. 5, 496 (1972)
47) Deam, R. T., Edwards, S. F.: Phil. Trans. Royal Soc. London, A-280, 317 (1976)
48) Candau, F., Rempp, P., Benoit, H.: Macromol. 5, 617 (1972)
49) Rietsch, F.: Thesis, Lille (1976)
50) Haeringer, A., Hild, G., Rempp, P., Benoit, H.: Makromol. Chem. 169, 249 (1973)
51) Candau, F., Strazielle, C., Benoit, H.: Europ. Polym. J. 12, 95 (1976)
52) Scholte, Th. G.: J. Polym. Sci. C-39, 281 (1972)
53) Scholte, Th. G.: J. Polym. Sci. A-2(8), 841 (1970)
54) Scholte, Th. G.: J. Polym. Sci. A-2(9), 1553 (1971)
55) Farnoux, B., Daoud, M., et al.: J. Phys. Let. 36, 135 (1975)
56) Daoud, M., Cotton, J. P., et al.: Macromol. 8, 804 (1975)
57) Treloar, L. R. G.: The physics of rubber elasticity. Oxford: Clarendon 1975
58) Borchard, W.: Diploma Thesis, T. H. Aachen (1962)
59) Rietsch, F., Froelich, D.: Polymer 16, 873 (1975)
60) Munch, J. P., Candau, S., Herz, J., Hild, G.: J. Phys., in press (1977)
61) Munch, J. P., Lemarechal, P., Candau, S., Herz, J.: to be published
62) Bastide, J., Picot, C.: to be published
63) Brault, A.: Thesis (3é. cycle), Lille (1976)
64) De Gennes, P. G.: J. Chem. Phys. 55, 572 (1971)
65) Hild, G., Froelich, D., Rempp, P., Benoit, H.: Makromol. Chem. 151, 59 (1972)
66) Weiss, P., Herz, J., Rempp, P., Gallot, Z., Benoit, H.: Markomol. Chem. 145, 105 (1971)
67) Rietsch, F., Daveloose, D., Froelich, D.: Polymer 17, 859 (1976)

Received May 25, 1977
W. Kern (editor)

NMR Approach to the Phase Structure of Linear Polyethylene

Ryozo Kitamaru and Fumitaka Horii

Institute for Chemical Research, Kyoto University, Uji, Kyoto, 611 Japan

The recent studies of the phase structure of linear polyethylene by refined NMR analyses are reviewed. The phase structure of the polymer in various crystalline forms, including bulk-crystals, solution-crystals and drawn fibers, is discussed in terms of different modes of molecular mobilities in a wide range of temperature.

Table of Contents

I. General Introduction

A group of long-chain molecules classified as crystalline polymers are not in thermodynamic equilibrium in the solid state. They have a variety of phase structures, depending strongly on the mode and conditions of their production. It is confirmed by X-ray
diffraction analysis that this group of polymers do generally contain a crystalline
phase or region in which molecules or molecular chains are regularly aligned in space
in a comparable manner to the crystals of monomeric substances. However, they
cannot by any means be fully crystalline, since the coexistence of a noncrystalline
content is conclusively evidenced by a variety of thermodynamic quantities[1] as well
as relaxation phenomena observed in mechanical and dielectric measurements[2, 3].
It is thus revealed that crystalline polymers have a phase structure, comprising crystalline and noncrystalline regions.

    With regard to the crystalline region, detailed information on, for example,
molecular alignment or the aggregative aspect of crystallites can be given by X-ray
diffraction or electron-microscopic observation. On the other hand, with regard to
the noncrystalline content, there has been no proper method to obtain reliable information. The noncrystalline content has been assumed to be in a undercooled
version of the state of the completely molten polymers at high temperatures. Otherwise, a special conformation of molecules, such as the regularly folded chain conformation, has sometimes been postulated for molecular chains, based on the lamellar
structure observed by electron microscopy[4–6]. However, electron-microscopic observation essentially gives us no information regarding an order of the molecules.
Such a special conformation of molecules cannot be assumed without reservation.
Nevertheless, the molecular conformation in the noncrystalline region is unlikely
to be the same in detail as that of the completely molten state of the polymers. Since
a molecular chain in the structure generally participates in both the crystalline and
the noncrystalline regions, molecular mobility in the noncrystalline region will be
more or less restricted by the presence of the crystalline region.

    On the other hand, since the absorption spectrum of nuclear magnetic resonance
(NMR) for substances in the solid state, though it is generally very broad, is decided
by a random fluctuation in local magnetic field caused by mutual interactions between nuclei, the spectrum for crystalline polymers may reflect their naked phase
structure. The random fluctuation of the local magnetic field is described by the
so-called correlation function, or more concretely by a correlation time which characterizes the correlation function. This correlation time directly corresponds to the
relaxation time in the mechanical and dielectric behaviors and reflects the molecular
mobility. Hence, pertinent analysis of the spectrum is expected to give us important
information on either region of the crystalline polymers, particularly on the noncrystalline region, which we have previously had no proper method for investigating.

    The theoretical absorption spectrum $A(\omega)$ under a constant static mainfield is
given as a function of the frequency of subfield by[7]

$$A(\omega) = (1/2\,\pi)\,\text{Re} \int_{-\infty}^{\infty} \exp\left[-\sigma_0^2\,\tau_c^2\,\{\exp\left(-|t|/\tau_c\right) + |t|/\tau_c - 1\}\right]$$
$$\times \exp\{-i\,(\omega-\omega_0)t\}\,dt \tag{1^a}$$

where σ_0^2 is the second moment of the line in the adiabatic state, ω is the angular frequency of the rotating field in NMR measurement, ω_0 is the resonance angular frequency for an isolated nucleus, t is the time elapsed since excitation, τ_c is the correlation time. If τ_c is short enough and $\sigma_0 \tau_c < 1^b$, Eq. (1) reduces to a Lorentzian with a small line-width of $(2/\sqrt{3})\sigma_0^2 \tau_c$ which is defined to be the difference in frequency between the extremes of slopes

$$A(\omega) = (1/\pi) \frac{\sigma_0^2 \tau_c}{(\sigma_0^2 \tau_c)^2 + (\omega - \omega_0)^2} \tag{2)^a}$$

In another extreme case, if τ_c becomes large enough, Eq. (1) reduces to a Gaussian with a large line-width of $2\sigma_0$

$$A(\omega) = (2\pi)^{-1/2} \sigma_0^{-1} \exp\left[-(\omega - \omega_0)^2 / 2\sigma_0^2\right]. \tag{3)^a}$$

For an intermediate value of τ_c the absorption spectrum takes a certain form between the Lorentzian and Gaussian. Thus, the value of τ_c decides the spectrum for substances. It is to be noted here that if τ_c becomes smaller than about 10^{-5} sec the spectrum generally becomes so narrow that a pronounced narrowing of the absorption line is abruptly observed.

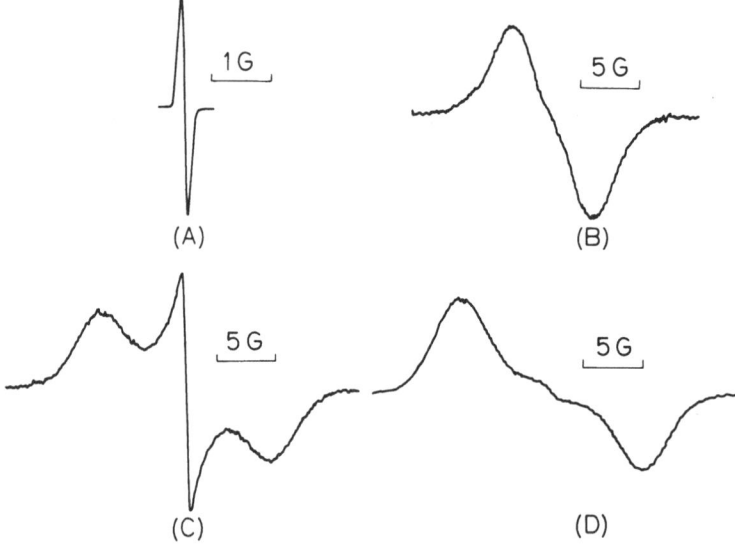

Fig. 1. Broad-line proton NMR spectra for some polymers. (A) polybutadiene at 20 °C, (B) polyethylene terephthalate at 20 °C, (C) polyethylene at 20 °C, (D) polyethylene at −150 °C

---

a   These equations can be also expressed as functions of the strength H of the main magnetic field by using a relation of $\omega = \gamma H$, were γ is the nuclear gyromagnetic ratio. Such expressions may be more useful for the usual broad-line NMR spectrometry in which the main field is slowly swept under a constant rotating subfield.

b   For proton magnetic resonance of linear polyethylene, since $\sigma_0^2 = 4.71 \times 10^8$ Hz^2 [8], this relation corresponds to $\tau_c < 4.6 \times 10^{-5}$ sec.

Figure 1 demonstrates the broad-line proton magnetic resonance spectra for some polymers. Here, first derivatives of absorption spectra are plotted against the main magnetic field in gausses (G). The spectrum for polybutadiene at 20 °C shows a Lorentzian-like line shape in differentiated form with a small line-width of about 0.2 G, whereas the spectrum for polyethylene terephthalate at 20 °C is very broad, with a line-width of about 7.3 G. These spectra well reflect the rubbery and glassy states of polybutadiene and polyethylene terephthalate, with small and large correlation times, respectively. On the other hand, the spectrum for linear polyethylene at −150 °C rather resembles that for polyethylene terephthalate at 20 °C and reflects its glassy state at −150 °C. However, the spectrum at 20 °C seems to be a superposition of narrow and broad components, with small and large correlation times, respectively. This implies that the polymer includes in the structure at least two different regions with distinctly different molecular mobilities, associated with shorter and longer correlation times.

In early studies of crystalline polymers, such a spectrum was somewhat arbitrarily divided into two parts, narrow and broad components, and their intensity fractions to the total spectrum (the area fractions to the total spectrum in integrated form) were defined to be mobile and immobile (or rigid) fractions, respectively. In such a case, the spectrum was divided into two parts by a straight line passing through the origin (straight line decomposition method[9]) or by assuming that the line shape of the broad component was approximately symmetrical about its maximum (symmetrical decomposition method[10, 11]), as shown in Fig. 2. If the broad component corresponds to the crystalline region, the rigid fraction should coincide with

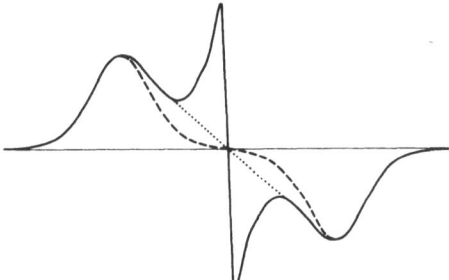

Fig. 2. Demonstrative two-component analyses for a broad-line spectrum; *dotted line:* straight line decomposition[9], *dashed line:* symmetrical decomposition[10, 11]

the crystallinity in a morphologic mean. However, it has been generally recognized for many polymers that the rigid fraction is appreciably larger than the crystallinity obtained by X-ray diffraction and density measurements. This discordance is thought to be caused by the fact that the noncrystalline region behaves in a mobile way only when the correlation time is shorter than 10^{-5} sec, and can contribute to the broad component if the correlation time is longer than that limit. Therefore, in order to examine the phase structure correctly, a more refined analysis of such spectra in relation to the correlation time for each phase is called for.

Recently Bergmann and Nawotki[12−15] have developed a method to decompose such spectra for polyethylene into three components: broad, intermediate, and narrow components. The methylene groups of the polymer were divided into three classes: rigid, hindered-rotational, and micro-Brownian mobile CH_2-groups, and these were considered to contribute to the broad, medium, and narrow components, respectively. They have certified by this new technique that the fraction of the broad component obtained corresponds closely to the crystalline fraction and that the noncrystalline region comprises two kinds of methylene groups. The one has a liquidlike mobility, the other has less mobility. Furthermore, they have argued that the relaxation processes, described as α-, β-, and γ-processes, that are observed in the dielectric and mechanical measurements, correspond to processes associated with a local molecular movement in the crystalline region, and micro-Brownian and limited molecular motions in the noncrystalline region, respectively. Although similar attempts to decompose the NMR spectra into multiple components have been reported[16−21], their method seems to be more reasonable and more widely applicable to different samples, granted that it is a little laborious. For oriented polyethylene samples Fischer et al.[22] decomposed the spectra into four parts: crystalline, oriented intermediate, unoriented intermediate, and liquidlike components, based on the method of Bergmann and Nawotki.

We have analyzed the broad-line NMR spectrum in a wide range of temperatures for linear polyethylene samples with different molecular weights, which were crystallized or processed in different modes, principally according to the three-component analysis developed by Bergmann and Nawotki[14] in connection with the phase structure of the samples. Although the phase structure of the samples examined varied over a very wide range, the NMR analysis could be achieved for all samples and very detailed information concerning the phase structure was obtained. This article mainly reviews this series of work.

II. Molten State

It is widely recognized[23] from the results of a variety of experimental methods, such as neutron scattering[24−33], small-angle X-ray scattering[34−37], light scattering[38], etc., that the configuration of molecular chains of bulk amorphous polymers is the same as in a θ-solvent, as predicted using the statistical theory developed by Flory[39]. Thus, in the bulk amorphous polymers, either in the glassy or rubbery state, no intermolecular order beyond the dimension of molecules is considered to exist. However, a conflicting argument is also presented, based mainly on an electron-microscopic observation[40, 41], that the structure is not homogeneous but inhomogeneities such as bundles of molecular chains in a microscopic order exist.

In NMR spectroscopy, if the structure is homogeneous and the molecular mobility is large enough, e.g., $\sigma_0 \tau_c < 1$, the absorption spectrum must reduce to a Lorentzian curve as discussed in Chapter I. In fact, it is observed that monomeric substances do form a Lorentzian in the melt or solutions; however, macromolecular substances hardly form such a single Lorentzian even in the melt[42, 43]. For example, Fig. 3 shows a NMR absorption spectrum for a molecular weight fraction of linear polyethylene in the melt at 147 °C[43]. Although the least-squares fitting by a single Lorentzian is tried, it is evident that the observed spectrum cannot be represented by a single Lorentzian. Gölz and Zachmann[19, 42, 44, 45] cited this fact as evidence of

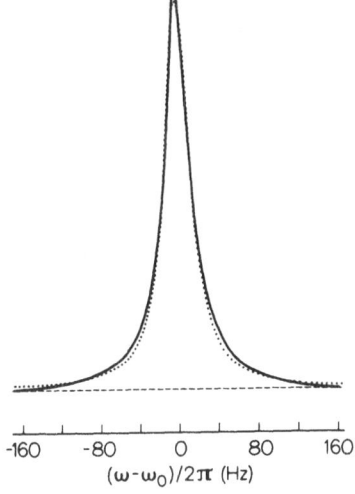

Fig. 3. Analysis of the high resolution NMR spectrum for a molecular weight fraction of linear polyethylene with $\bar{M}\eta = 11\,900$ by one Lorentzian curve with $\tau_c = 3.33 \times 10^{-8}$ sec and $\sigma_0^2 = 4.71 \times 10^8$ Hz^2. The solid and the dotted lines indicate the observed and the calculated spectra, respectively[43]

$(\omega - \omega_0)/2\pi$ (Hz)

structural inhomogeneity in the molten polymer. Since such a spectrum could be analyzed into two or more Lorentzian lines, they argued that inhomogeneities such as bundles of parallel molecular chains were formed. However, this deviation from a Lorentzian line could not certify the presence of inhomogeneities. The molecular motion for long-chain molecules is thought to be rather complicated. As is well known, in the mechanical and dielectric relaxation times for polymeric substances very wide distributions are always assumed. Therefore a distribution should also be assumed for the correlation time which describes the random fluctuation of local magnetic field due to the molecular motions[46, 47].

Thus, Eq. (2) must be rewritten as follows

$$A(\omega) = (1/\pi) \int_{-\infty}^{\infty} \frac{\sigma_0^2 \tau_c}{(\sigma_0^2 \tau_c)^2 + (\omega - \omega_0)^2} \, I(\tau_c) d\ln\tau_c \tag{4}$$

$$\int_{-\infty}^{\infty} I(\tau_c) d\ln\tau_c = 1. \tag{5}$$

Here, $I(\tau_c)$ is the correlation spectrum, $I(\tau_c)d\ln\tau_c$ is the probability that the logarithm of an arbitrarily chosen correlation time has a value between $\ln\tau_c$ and $\ln\tau_c + d\ln\tau_c$. If a rectangular distribution for $\ln\tau_c$ is assumed in a range between τ_{ca} and τ_{cb}:

$$I(\tau_c) = \begin{cases} 1/\ln(\tau_{cb}/\tau_{ca}) & \text{for } \tau_{ca} \leq \tau_c \leq \tau_{cb} \\ \\ 0 & \text{otherwise} \end{cases} \tag{6}$$

Then, Eq. (4) reduces to

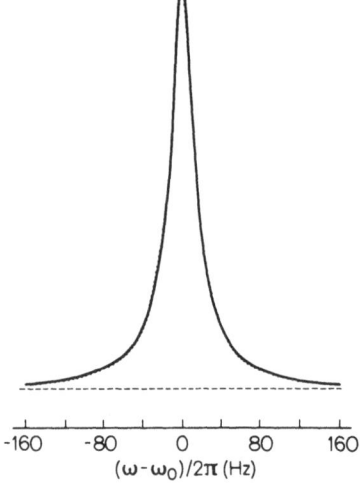

Fig. 4. Analysis of the high resolution NMR spectrum for a molecular weight fraction with $\bar{M}\eta = 11\,900$ according to Eq. (7) with $\tau_{ca} = 1.81 \times 10^{-8}$ sec, $\tau_{cb} = 8.42 \times 10^{-8}$ sec, and $\sigma_0^2 = 4.71 \times 10^8$ Hz^2. The solid and the dotted lines indicate the observed and the calculated spectra, respectively[43]

$$
A(\omega) = \begin{cases} \dfrac{1}{\pi(\omega-\omega_0)\ln(\tau_{cb}/\tau_{ca})}\left(\tan^{-1}\dfrac{\sigma_0^2 \tau_{cb}}{\omega-\omega_0} - \tan^{-1}\dfrac{\sigma_0^2 \tau_{ca}}{\omega-\omega_0}\right) & \text{for } \omega \neq \omega_0 \\[2em] \dfrac{1}{\pi\ln(\tau_{cb}/\tau_{ca})} \cdot \dfrac{\tau_{cb}-\tau_{ca}}{\sigma_0^2 \tau_{ca}\tau_{cb}} & \text{for } \omega = \omega_0 \end{cases} \tag{7}
$$

In Fig. 4 least-squares fitting with this equation to the experimentally observed spectrum is shown. The dotted line expresses Eq. (7) with $\tau_{ca} = 1.81 \times 10^{-8}$, $\tau_{cb} = 8.42 \times 10^{-8}$ sec (Here, τ_0^2 is assumed to be 4.71×10^8 Hz^2 [8])). The observed and calculated spectra superpose with no appreciable deviation. Similar excellent agreement between observed and calculated spectra was obtained for many samples with different molecular weights. Thus, the deviation of spectra for polyethylene in the melt from a single Lorentzian is understood by the distribution of the correlation time.

In conclusion, NMR spectroscopy on polyethylene in the melt implies the existence of a variety of segmental motions characteristic of long-chain molecules, but does not support the argument that the structure is not homogeneous.

III. Analysis of Broad-Line NMR Spectrum

As pointed out in Chapter I, the NMR absorption spectrum for polymers in the solid state is generally very broad. In such cases the NMR is usually observed as the so-called broad-line spectrum. In this, the resonance is recorded by slowly sweeping the main static magnetic field H′ modulated with a small amplitude and frequency under a constant high-frequency subfield rotating perpendicularly to the main field. Figure 5 shows schematically the principle of the measurement for the broad-line NMR

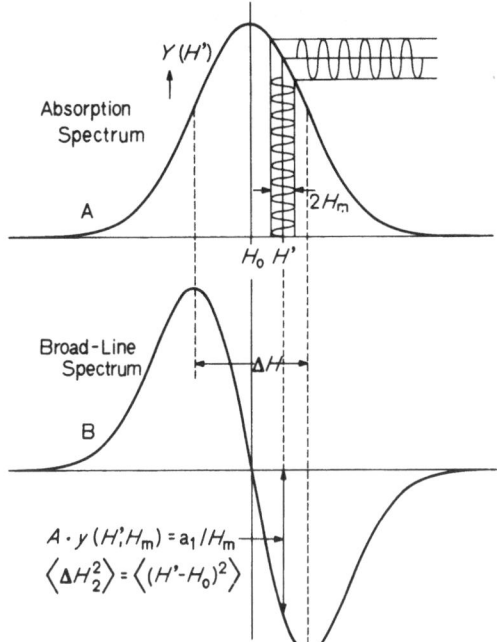

Fig. 5. Relation between a NMR absorption spectrum and its first derivative spectrum (broad-line spectrum). See the text for explanations

spectrum. Let the curve $A[Y(H')]$ be an absorption line obtained as a function of field strength under a constant subfield, then the broad-line spectrum is obtained as the curve B when the measurement is made with a modulation amplitude of H_m. In the figure the definition of the line-width ΔH and the second moment $\langle \Delta H_2^2 \rangle$ are given. ΔH is defined to be a field difference between two extremes in the absorption line or the broad-line spectrum (the extremes in slopes or in detected values, respectively). The second moment $\langle \Delta H_2^2 \rangle$ is defined to be the average of the squares of the deviation of the field from the center of the resonance H_0,

$$\langle \Delta H_2^2 \rangle = \langle (H' - H_0)^2 \rangle. \tag{8}$$

It is evident that both ΔH and $\langle \Delta H_2^2 \rangle$ are obtained from either the absorption or the broad-line spectrum.

When the main field is modulated with an angular velocity ω_m, the absorption will be produced as a function of $H' + H_m \cos \omega_m t$. Expanding this modulated absorption into the Fourier series, we obtain

$$Y(H' + H_m \cos \phi) = a_0/2 + \sum_{k=1}^{\infty} a_k \cos k\phi \tag{9}$$

with $\phi = \omega_m t$. Here t designates time, and the coefficients in the equation are given as

$$a_k(H') = 1/\pi \int_{-\pi}^{\pi} Y(H' + H_m \cos \phi) \cos k\phi\, d\phi \qquad k = 0, 1, 2 \cdots \tag{10}$$

When a component of the modulated absorption wave which synchronizes with the kth harmonic of the modulation frequency is detected by a phase-sensitive detector, the signal obtained is proportional to the coefficient of the kth term $a_k(H')$.

On the other hand, the Taylor expansion of Y gives the coefficient a_k with de Moivre's theorem in the form[48, 49]

$$a_k(H') = \sum_{m=0}^{\infty} H_m^{k+2m} \alpha_{k+2m,k} Y^{(k+2m)}(H') \qquad (11)$$

where

$$\alpha_{i,j} = \begin{cases} 2^{1-i} \left\{ \left(\frac{i+j}{2}\right)! \left(\frac{i-j}{2}\right)! \right\}^{-1} & \text{for } i \geq j \text{ and } (i-j) \text{ even,} \\ \\ 0 & \text{for } i < j \text{ and } (i-j) \text{ odd.} \end{cases}$$

Here $Y^{(k+2m)}(H')$ refers to the $(k+2m)$th derivative of $Y(H')$ with respect of H'. For low modulation amplitudes Eq. (11) reduces to

$$a_k(H') = Q_k H_m^k Y^{(k)}(H')$$

where

$$Q_k = \begin{cases} 2^{1-k}/k! & (k \neq 0) \\ \\ 1 & (k=0) \end{cases} \qquad (12)$$

Therefore if the modulation amplitude is small enough, the output signal of the phase-sensitive detector can be considered to be proportional to the kth derivative of the absorption spectrum $Y(H')$. In usual broad-line NMR spectroscopy the first derivative is recorded by a phase-sensitive detector employing detection at the first harmonic[c]. However, if larger amplitudes are employed for the measurement, the spectrum obtained will be appreciably distorted from the true first-order differential. Larger modulation amplitude, of course, gives a larger signal for the broad-line NMR measurement, but in such a case a deviation from the true differential must be considered. We will discuss this matter later.

In the following chapters we will discuss the phase structure of linear polyethylene samples in the solid state by analyzing the broad-line proton NMR spectrum by a technique developed by Bergmann and Nawotki[12-14], decomposing the spectrum into three parts: broad, medium, and narrow components. We first therefore review the three-component analysis used in this series of work.

---

[c]   Slonim *et al.*[50, 51] recorded the second derivative of the absorption spectrum for some polymers.

The first derivative of a broad-line spectrum can be described in the three terms as a function of the strength of the main magnetic field in the measurement:

$$y(H) = w_b y_b (H, \beta_b) + w_m y_m (H, \beta_m) + w_n y_n (H, \beta_n) \tag{13}$$

with

$$w_b + w_m + w_n = 1$$

Here, H can be conveniently expressed as the deviation of the field from the center of the resonance in gauss units: $H = H' - H_0$; y_b, y_m, and y_n are the elementary spectra of the broad, the medium, and the narrow components, respectively. These are considered to be contributed from protons belonging to the crystalline region, and hindered-rotational and micro-Brownian mobile methylene groups in the amorphous region, respectively. β_b, β_m, and β_n determine the line-width and breadth of the respective elementary spectra; w_b, w_m, and w_n designate the respective mass fractions. Each elementary spectrum is normalized as

$$\int_{-\infty}^{\infty} \int_{-\infty}^{H} y_i(H, \beta_i)dH = 2, \qquad i = b, m, \text{ and } n. \tag{14}$$

The parameters in Eq. (13) are to be determined so as to minimize the sum Φ of the squares of the difference between the observed and calculated spectra over the full range of H:

$$\Phi = \sum_j \{ y_{obs} (H_j) - A y (H_j) \}^2 \tag{15}$$

Here A is a parameter which adjusts the amplitude of the spectrometry. The success for this three-component analysis relies on adequate choice for the elementary spectra. Hence, we review briefly the elementary spectra used for this series of work.
Broad Component. The elementary spectra for the broad components were obtained by changing the line-width of the spectrum at $-150\,^{\circ}C$ of the HNO_3-treated polyethylene with a very high crystallinity, but keeping the shape unchanged. When the spectrum $y_s(H)$ at $-150\,^{\circ}C$ for the highly crystalline polymer has the line-width ΔH_s, an elementary spectrum $y_b (H)$ with a line-width ΔH_b can be constructed as

$$y_b (H) = \chi y_s (\gamma H), \qquad \gamma = \Delta H_s / \Delta H_b \tag{16}$$

Here χ designates the normalization factor.
Medium Component. The elementary spectra for the medium component were derived by Bergmann and Nawotki[12, 13] on the basis of a spectrum calculated by Gutowsky and Pake[52] for paired protons such as in 1,2-dichloroethane with hindered rotation around the C–C bond.

$$y_m(H, \beta_m) = \frac{1}{4(6\pi)^{1/2}\,\alpha\beta_m^3} \int_{-\alpha}^{2\alpha} \frac{(2H+h)\exp\left\{-\dfrac{(2H+h)^2}{8\beta_m^2}\right\} + (2H-h)\exp\left\{-\dfrac{(2H-h)^2}{8\beta_m^2}\right\}}{(1+h/\alpha)^{1/2}}\,dh$$

(17)

with $\alpha = (3/2)\,\mu r^{-3} = 4.0 \pm 0.1$ G

Here r is the distance between the paired protons, μ is the magnetic moment of the proton. However, the molecular mobility in the amorphous region in the polymer where the rotation around C–C bonds is hindered is not so simple as expected from Eq. (17). The equation cannot adequately reproduce the medium component of the spectrum. It is reported by Bergmann and Nawotki[14] that for actual spectra for polymers the following expression reproduces more adequately the medium component:

$$y_m(H, \beta_{mg}, \beta_{ml}) = N\frac{\partial}{\partial H}\exp\left(-\frac{H^2}{2\beta_{mg}^2}\right) \cdot \frac{\beta_{ml}^2}{\beta_{ml}^2 + H^2}$$

(18)

Here N designates the normalization factor. Clearly this equation in integrated form is the product of Gaussian and Lorentzian distribution functions; β_{mg} and β_{ml} define the line-widths of the two components, respectively. Here, the former represents Eq. (17) to a sufficient approximation for $\beta_m \geq 2$ G and the latter was introduced to express the coupled rotational and/or the translational motion of proton pairs in the polymer, discussed by Pechhold[53].

Narrow Component. As discussed in Chapter II, the absorption spectrum for polyethylene cannot be described by a single Lorentzian even in the molten state. However, the deviation from one Lorentzian is not enhanced for well-fractionated samples in the melt and, furthermore, becomes negligible as the temperature decreases[42]. Accordingly, the differential form of a Lorentzian distribution can be used for the elementary spectrum of the narrow component:

$$y_n(H, \beta_n) = 4\pi^{-1}\beta_n^{-3} H/(1 + \beta_n^{-2} H^2)^2$$

(19)

or in the integrated form,

$$Y_n(H, \beta_n) = 2(\pi\beta_n)^{-1}/(1 + \beta_n^{-2} H^2)$$

(20)

Here, β_n is related to the line-width ΔH_n by $\beta_n = (\sqrt{3}/2)\Delta H_n$ and

$$\int_{-\infty}^{\infty} Y_n(H, \beta_n)\,dH = 2$$

However, since the line-width of the narrow component is generally rather small for most samples studied, the line shape is distorted by the amplitude of the modulation H_m, as pointed out already. It is well known that if the condition $H_m \leq \Delta H/5$ is fulfilled, distortions due to the H_m are negligible. Since this condition is generally not fulfilled for the narrow component, the distortion due to the H_m must be taken into account. Letting Y in Eq. (10) be $2(\pi \beta_n)^{-1}/\{1 + \beta_n^{-2}(H + H_m \cos \phi)^2\}$ we obtain the broad-line spectrum as a function of H, β_n, and H_m.

$$y_n(H, \beta_n, H_m) = a_1(H)/H_m$$

$$= \frac{2}{\pi^2 \beta_n H_m} \int_{-\pi}^{\pi} \frac{\cos \phi \, d\phi}{1 + \beta_n^{-2}(H + H_m \cos \phi)^2} \tag{21}$$

The finite integral appearing in the right-hand side of the equation has been solved by many investigators[13, 18, 54–57]. Substituting $\tan(\phi/2) = x$, we obtain

$$y_n(H, \beta_n, H_m) = \frac{2}{\pi^2 \beta_n H_m} \int_{-\infty}^{\infty} \frac{(1 - x^2) \, dx}{\xi x^4 + 2\eta x^2 + \zeta} \tag{22}$$

with

$$\xi = \beta_n^{-2}(H - H_m)^2 + 1$$
$$\eta = \beta_n^{-2}(H^2 - H_m^2) + 1$$
$$\zeta = \beta_n^{-2}(H + H_m)^2 + 1$$

Then the finite integration is easily obtained with the use of the complex integration as

$$y_n(H, \beta_n, H_m) = \frac{2^{3/2}}{\pi \beta_n H_m} \cdot \frac{\zeta^{1/2} - \xi^{1/2}}{(\xi \zeta)^{1/2} \{(\xi \zeta)^{1/2} + \eta\}^{1/2}} \tag{23}$$

Here $y_n(H, \beta_n, H_m)$ is normalized as

$$\lim_{H_m \to 0} \int_{-\infty}^{\infty} \int_{-\infty}^{H} y_n(H, \beta_n, H_m) \, dH = 2$$

Least Squares Estimation. For the three-component analysis employed in this series of work, Eq. (11) should be rewritten as

$$y(H, H_m) = w_b y_b(H, \Delta H_b) + w_m y_m(H, \beta_{mg}, \beta_{ml}) + w_n y_n(H, \beta_n, H_m) \tag{24}$$

Here $y_b(H, \Delta H_b)$, $y_m(H, \beta_{mg}, \beta_{ml})$, and $y_n(H, \beta_n, H_m)$ are given by Eqs. (16), (18), and (23), respectively. The parameters ΔH_b, β_{mg}, β_{ml}, β_n, two of w_b, w_m, w_n, and A in Eqs. (24) and (15) are to be determined so as to minimize the sum of squares of the difference between the observed and calculated spectra over the full range of field where the resonance appears. This least-squares calculation was carried out according to the so-called simplex method[d] which was proposed by Spendley et al.[59]

---

[d] Instead of the simplex method, any algorithm[58] for least squares estimation of nonlinear parameters may be used.

and developed by Nelder and Mead[60]. Since the peak of the broad component was clearly recognized and the line-width could be determined for most spectra examined, the parameter ΔH_b was equalized to the line-width or fixed at a value slightly larger than it by $0.1 - 0.3$ G and the remaining six parameters were allowed to float until the best fit was obtained according to the algorithm. Thus the optimum set of the parameters in Eq. (24) was determined for each observed spectrum and the mass fractions w_b, w_m, w_n as well as the line-widths or the second moments of the three components were discussed in connection with the phase structure of samples.

The three-component analysis described here can be used universally for many crystalline polymers such as polyoxymethylene[15], polyethylene oxide[15], polyvinylidene chloride[15], nylon-6 [15, 61], polypropylene[15], etc., if a proper elementary spectrum is used for the broad component, depending on the polymer sort. The spectrum for a polymer at a very low temperature, where any local motion of chains is generally hindered, does not depend appreciably on the crystallinity, but coincides with that of the completely crystalline material to a sufficient degree of approximation. Therefore, for the elementary spectrum for the broad component, a spectrum observed at a very low temperature for the respective polymer or its highly crystalline sample, if available, can be universally used.

IV. Samples Crystallized from the Melt

A. Introduction

In Figure 6 the density of the crystal unit-cell, estimated at room temperature by an X-ray diffraction technique, for linear polyethylene samples crystallized from the melt (hereafter designated bulk-crystallized samples or bulk-crystals) is plotted against the measured macroscopic density[62]. The figure includes data for samples which differ not only in the molecular weight but also in the mode of crystallization. Some of them were isothermally crystallized at high temperature near the melting point and some were rapidly cooled from the melt. Nevertheless, the density of the unit-cell actually stays unchanged within 0.999 ± 0.003 while the macroscopic density changes over a very wide range from 0.917 to 0.994. These results, together with

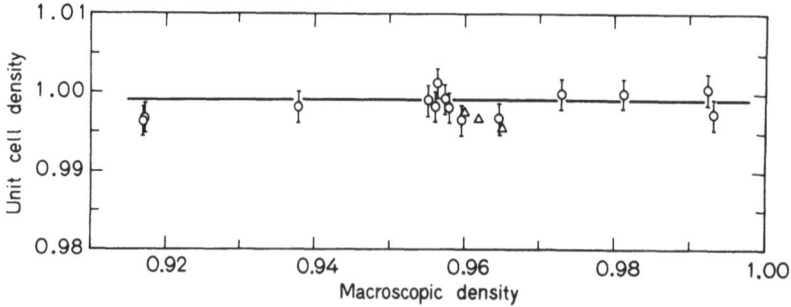

Fig. 6. Unit-cell density vs. macroscopic density for bulk-crystallized linear polyethylene[62]

the fact that samples with smaller macroscopic densities are always associated with smaller enthalpies of fusion[63], clearly indicate that the bulk-crystals comprise a crystalline phase with a well-defined density as well as a noncrystalline material with a lower density.

On the other hand, it has been recognized by electron-microscopic oberserva-tions[64, 65] that the bulk-crystals generally form lamellalike crystallites with a limited thickness. The molecular chain axes in the crystallites are preferentially oriented perpendicular to the basal plane of the lamella crystallites. Therefore, the noncrys-talline material is considered to be composed of molecular chains which are excluded from the basal plane of the lamella crystallites and form an interfacial region of them. Furthermore, the relative ratio of the extended molecular chain length to the lamella thickness can change from about unity to more than twenty according to the molec-ular weight[65]. Accordingly, the mode of participation of the molecular chain in the lamella crystallites as well as the molecular conformation of the noncrystalline ma-terial are thought to vary over a very wide range.

The phase structure of the bulk-crystals must be, of course, characteristic of this mode of crystallization, but the detail will be different, depending strongly on the molecular weight. In this chapter we review the NMR analysis[66, 67] for this kind of samples covering a very wide range of molecular weight in view of the change in the phase structure.

We deal with molecular weight fractions of the polymer which were crystallized from the melt under well-controlled conditions. The average molecular weight of samples covers a very wide range from 1 850 to 3.4×10^6 and the NMR measure-ments were carried out at temperatures from $-150°$ to $+60°$ or $+100\ °C$. A sample with an average molecular weight of 1 850 was slowly cooled to room temperature from the melt but the other samples were crystallized isothermally at 130 °C from the melt for 2 to 4 weeks, depending on the molecular weight, and cooled to room temperature over a period of a week. The three-component analysis of the NMR spectrum was carried out according to the method described in Chapter III.

B. Molecular Weight Dependence of the Phase Structure at Room Temperature

Figure 7 demonstrates the room temperature spectra and the three-component analysis for samples with molecular weights of $n-C_{44}H_{90}$, 20 500, 90 000, and 3.4×10^6. Although the molecular weights vary over a very wide range, the composite curves of the three components coincide satisfactorily with the observed spectra. The spec-trum for the sample with the lowest molecular weight predominantly comprises the broad component with minor medium and narrow components, rather resembling that for $n-C_{44}H_{90}$. As the molecular weight increases, the medium and narrow com-ponents increase. This clearly shows that the phase structure of samples depends greatly on the molecular weight. The sample with the lowest molecular weight mainly comprises only the crystalline region but the amorphous contents corresponding to the medium and narrow components appear with increasing molecular weight.

Figure 8 shows the relationship between the mass fraction w_b of the broad com-ponent and the crystallinity $(1-\lambda)_d$ obtained from the density measurements for

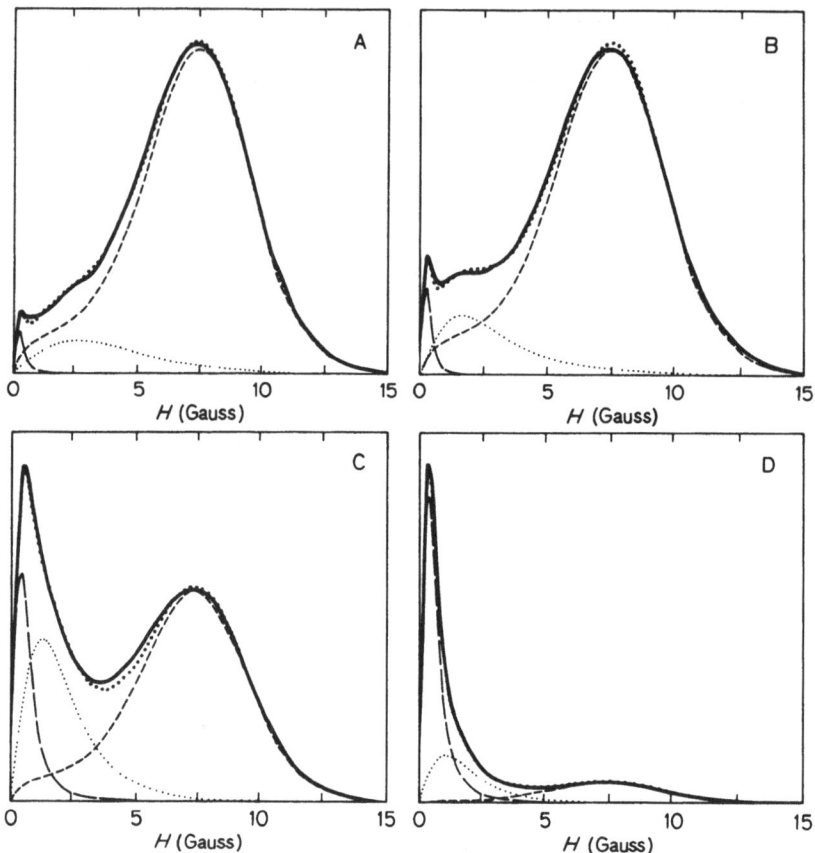

Fig. 7. Three-component analysis for bulk-crystals. Dashed, dotted, and broken lines indicate narrow, medium, and broad components, respectively. Thick dotted and solid lines indicate the composite curve for the three components and the experimental spectrum, respectively.
(A) n-$C_{44}H_{90}$, (B) $\bar{M}\eta$ = 20500, (C) $\bar{M}\eta$ = 90000, (D) $\bar{M}\eta$ = 3.4 x 10^6

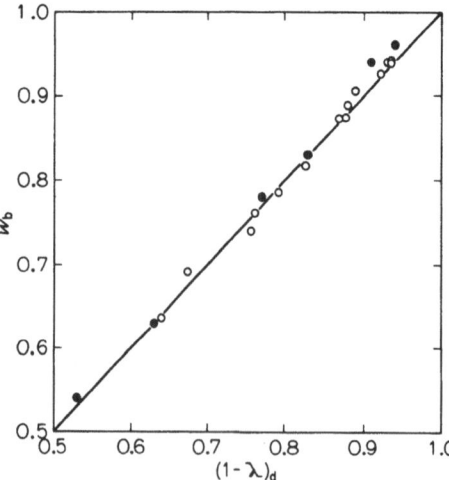

Fig. 8. Mass fraction of broad component vs. crystallinity obtained from density measurements[66]. Solid circles indicate data of Mandelkern et al.[68, 69]. Straight line indicates the relation, $(1-\lambda)_d = w_b$

samples covering the very wide range of molecular weight. The figure also includes data of Mandelkern and others[68, 69], obtained by a somewhat different technique[e], of the three-component analysis for samples which were crystallized in the same manner as in this series of work. It is evident that the mass fraction of the broad component coincides well with the degree of crystallinity calculated from the density over the very wide range. The degree of crystallinity from the density is known to be in good agreement also with that estimated from wide-angle X-ray diffraction or infrared absorption for polyethylene samples[70-72]. It is therefore apparent that the mass fraction of the broad component obtained from NMR analysis at room temperature represents well the morphologic crystalline fraction for the well-crystallized samples.

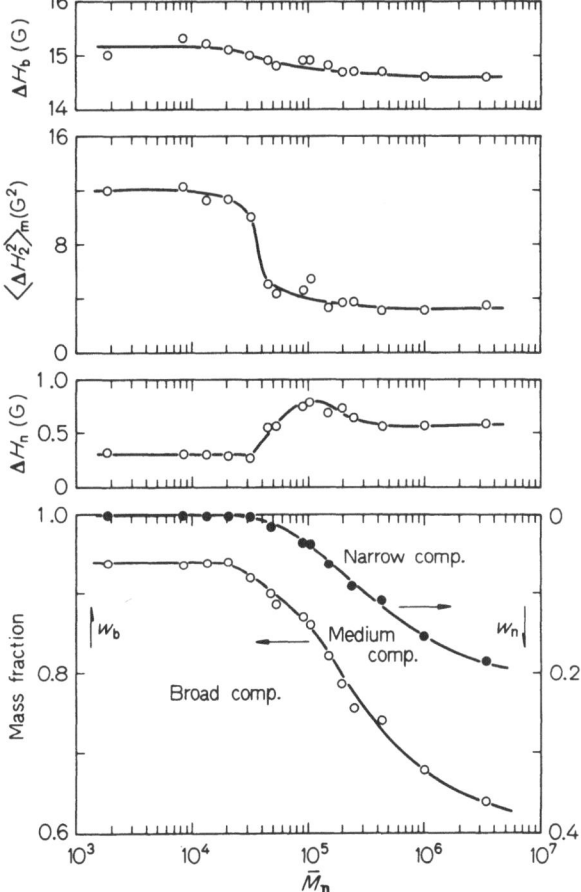

Fig. 9. Mass fractions, linewidths or second moment of three components for bulk-crystals as a function of molecular weight using a logarithmic scale. The same data were reported in Ref.[66] using an ordinary scale of molecular weight

---

[e] In their technique the least-squares calculation was carried out by equalizing the peak position of the narrow and broad components with those of the observed spectrum and using an elementary spectrum of the medium component which was obtained by modifying Pake's equation, i.e., Eq. (17). Independent of the analysis technique, the same result was obtained so far as concerned the mass fraction of the broad component.

The mass fractions and the line-widths or second moment of the three components are plotted against the average molecular weight in Fig. 9. It is seen that, although the narrow component is negligible for samples of molecular weights less than 31 800, it appears clearly at a molecular weight of 44 900 and increases as the molecular weight increases. Accompanying the significant appearance of the narrow component, its line-width abruptly increases and reaches an asymptotic level after going through a maximum.

The mass fraction of the medium component is as small as 0.05–0.08 for samples which are actually devoid of the narrow component but it gradually increases with the appearance of the narrow component and reaches a rather high level for samples with higher molecular weights. The second moment of the medium component is rather large for samples with lower molecular weights but it abruptly decreases when the narrow component appears and gradually decreases with increasing molecular weight. The mass fraction of the broad component, which represents well the degree of crystallinity as mentioned above, gradually decreases from a very high level ($w_b \simeq 0.94$) to a rather low level ($w_b \simeq 0.64$) as the molecular weight changes from 2000 to 3.4×10^6. The line-width decreases slightly (from 15.2 to 14.6 G) with increasing molecular weight.

The above-mentioned enhanced dependence of the three component analysis on the molecular weight must reflect great dependence of the phase structure on the molecular weight. The broad, medium, and narrow components are considered to reflect contributions from protons in the rigid and hindered-rotational methylene groups, and in the micro-Brownian mobile methylene groups together with methyl end-groups, respectively. Those components will correspond to crystalline, interfacial, and interzonal regions in the lamellar structure for samples crystallized in this mode of crystallization independent of the molecular weight. We next consider the phase structure of samples by the three-component analysis, using the ratio ζ/x of the lamellar thickness ζ to the extended molecular chain length x reported by Mandelkern and others[65] for samples which were crystallized under the same conditions as in this series of work.

1. Samples with Molecular Weights less than about 30000

Samples in this molecular weight range will be composed of the crystalline and interfacial regions but actually no interzonal region with liquidlike character as suggested by the negligible values of w_n. The interfacial region evidently exists to the extent of w_m 0.05–0.08 but the highest values of the second moment $\langle \Delta H_2^2 \rangle_m$ suggest that the molecular mobility in this region is rather strongly restricted. Within this molecular weight range the average ratio of the lamellar thickness to the extended molecular chain length, ζ/x, is believed to change from unity to about 1/4 [65]. In the structure of n-paraffins all methyl end-groups are paired in the crystal interface as depicted schematically in Fig. 10 (A). The structure of samples with lower molecular weights such that $x \simeq \zeta$ can be expressed by the so-called unpeeled crystal model depicted schematically in Fig. 10(B), where such pairing of the methyl end-groups will be lost due to the distribution of molecular weight and some methylene sequences neighboring the end-groups will be excluded from the crystalline region.

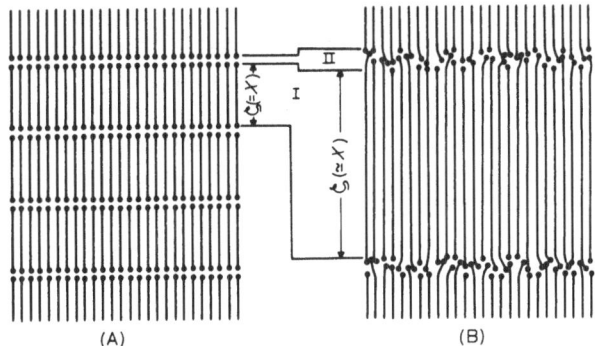

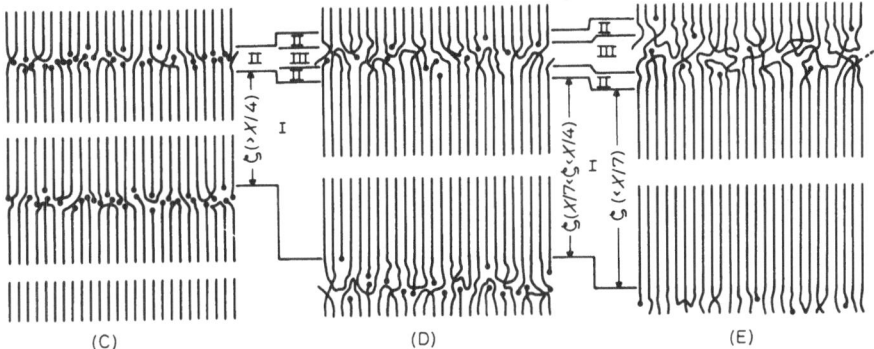

Fig. 10. Schematic structure models of the bulk-crystallized polyethylene samples. I, II, and III indicate the crystalline, interfacial, and interzonal regions, respectively. Models A, B, C, D, and E express the molecular crystal, unpeeled crystal, disheveled unpeeled crystal, and lamellar crystals for medium and large molecular weight samples, respectively[66]. ζ and x designate the lamellar thickness and the extended molecular chain length, respectively

A molecular chain participates only once in a lamellar crystallite when $x \simeq \zeta$. As ζ/x decreases from unity with increasing molecular weight, molecular chains that penetrate several lamellar crystallites or participate repeatedly in a lamellar crystallite will begin to appear as schematically shown by the disordered unpeeled crystal model in Fig. 10(C). Whether molecular chains that are excluded from the crystalline region and form the interfacial region participate again in the crystalline region or not, the interfacial region will be composed of relatively short sequences of methylene groups. The mobility of such short methylene sequences is thought to be strongly restricted by the crystalline region, resulting in the highest value of $\langle H_2^2 \rangle_m$.

2. Samples with Molecular Weights of 45000–100000

Within this molecular weight range, ζ/x varies from 1/4 to 1/7 [65]. As the molecular weight increases, the number of molecular chains that penetrate many lamellar crystallites or participate repeatedly in a crystallite will increase, accompanying a pronounced decrease in the degree of crystallinity. This will result in a loss in the restriction on mobility of noncrystalline molecular chains caused by crystalline regions. As a result, an interzonal region, in which molecular chains have a higher mobility because of the micro-Brownian movement of chains, will be produced in the center of the noncrystalline region as depicted schematically in Fig. 10 (D).

This will be reflected in the spectrum by a gradual increase in w_m and an abrupt decrease in $\langle \Delta H_2^2 \rangle_m$, accompanying the significant appearance of the narrow component. Within this molecular weight range, the narrow component will be contributed not only by protons in methyl end-groups and their neighboring methylene groups, but also by protons belonging to methylene groups in the interzonal region. Since the latter protons have lower mobility than the former, their contribution to the spectra will result in a pronounced increase in the ΔH_n, so that the ΔH_n indicates a clear maximum within this molecular weight range.

In compliance with the structure that gives the interzonal region with a liquidlike character in addition to the interfacial region, the samples begin to show the toughness characteristic of typical high polymers. In comparison with the n-tetratetracontane and polyethylene samples with molecular weights smaller than 31 800, the samples in this molecular weight range are not so brittle and cannot easily be fractured.

3. Samples with Molecular Weights of $100000 - 3.4 \times 10^6$

Above a molecular weight of 100000, ζ/x becomes progressively much smaller than $1/7$ [65]. Accordingly, accompanying the increase of the mass fraction of the interfacial region, the interzonal region will progressively increase and the liquidlike character of this region will become pronounced. This is reflected in the spectrum by a progressive decrease in ΔH_n and increase in w_m and w_n. Samples of larger molecular weight have a lamellar crystalline structure which is composed of the crystalline, interfacial, and interzonal regions that can be represented by the schematic diagram in Fig. 10 (E). This schematic model is substantially the same as Mandelkern[73] depicted three-dimensionally.

As discussed above, the interior structure of the noncrystalline regions of the samples with different molecular weights has been elucidated in detail by the spectrum analysis. On the other hand, Mandelkern et al.[65] have found that the interfacial free energy σ_{ec} in bulk-crystals is not constant, but increases with increasing molecular weight. This result represents a systematic change in the structure of the interfacial region as a function of the molecular weight, which can be schematically depicted by the models shown in Fig. 10. The NMR analysis in this work is not only consistent with their conclusion but also gives us more detailed information on the noncrystalline regions.

C. Temperature Dependence of the Phase Structure

We have known by examining the room temperature spectrum of the bulk-crystals that the phase structure depends strongly on the molecular weight but it is generally composed of the crystalline, interfacial, and interzonal regions, of which molecular mobilities differ with each other. The temperature dependency of the phase structure also differ to great extents depending on the molecular weight[67]. We discuss this problem separately dividing samples into three regions of molecular weight.

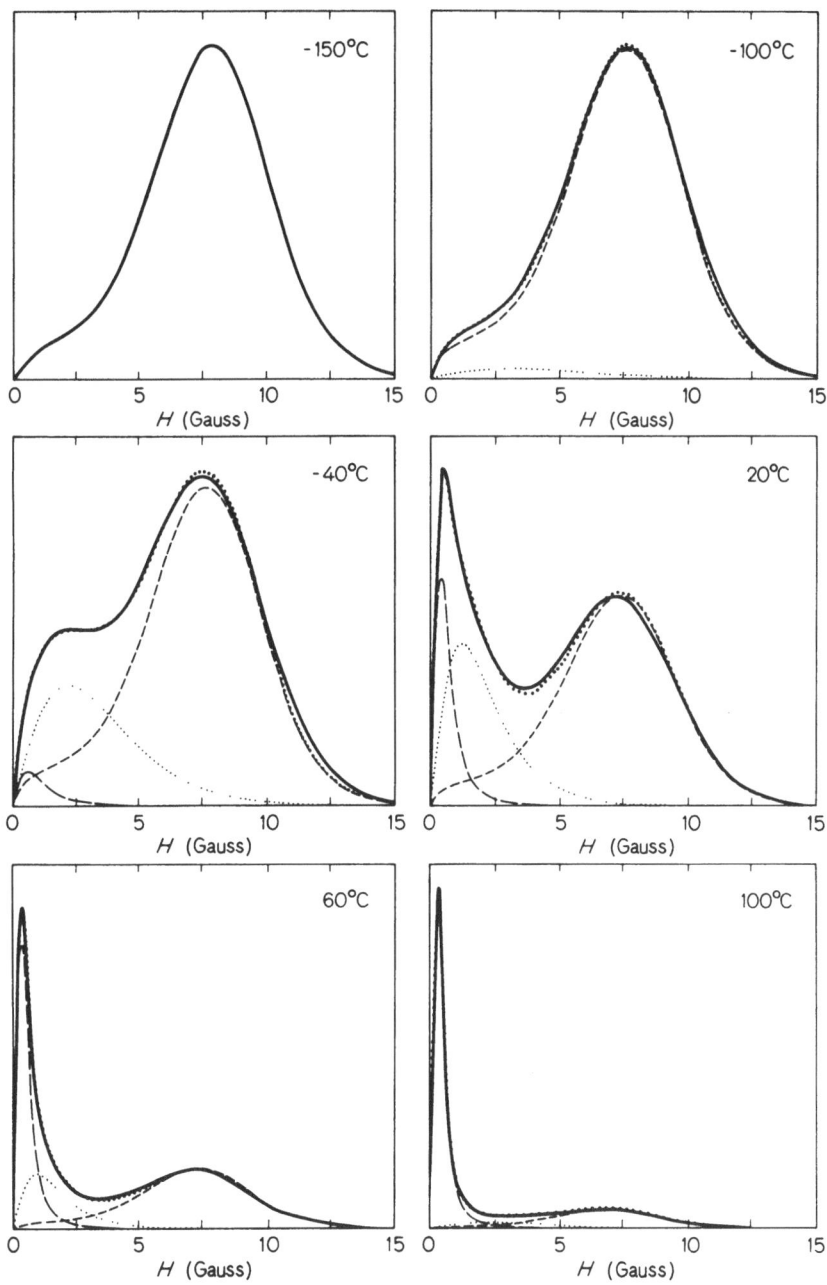

Fig. 11. Three-component analysis for a bulk-crystal with $\bar{M}\eta$ = 90000 over a wide range of temperatures. Dashed, dotted, and broken lines indicate narrow, medium, and broad components, respectively. Thick dotted and solid lines indicate the composite curve for the three components and the experimental spectrum, respectively

1. Sample with a Medium Molecular Weight

Figure 11 illustrates the three-component analysis of spectra for a sample with a molecular weight of 90000 at different temperatures from −150° to + 100 °C. At −150 °C the spectrum comprises only the broad component, with no medium and narrow components. It approximately coincides with an elementary spectrum for the broad component which was obtained by changing the line-width of the spectrum at −150 °C for the HNO_3-treated sample with a very high crystallinity. The medium component is recognized for the spectrum at −100 °C in addition to the broad component and the narrow component is recognized at temperature above −40 °C. Whether all of the three components appear in the spectrum or not, it is seen that the composite curves coincide with the spectra observed at all temperatures. These results not only indicate the adequacy of the three-component analysis for the spectrum of the polymer at different temperatures, but also suggest great temperature dependence of the molecular mobility in each phase.

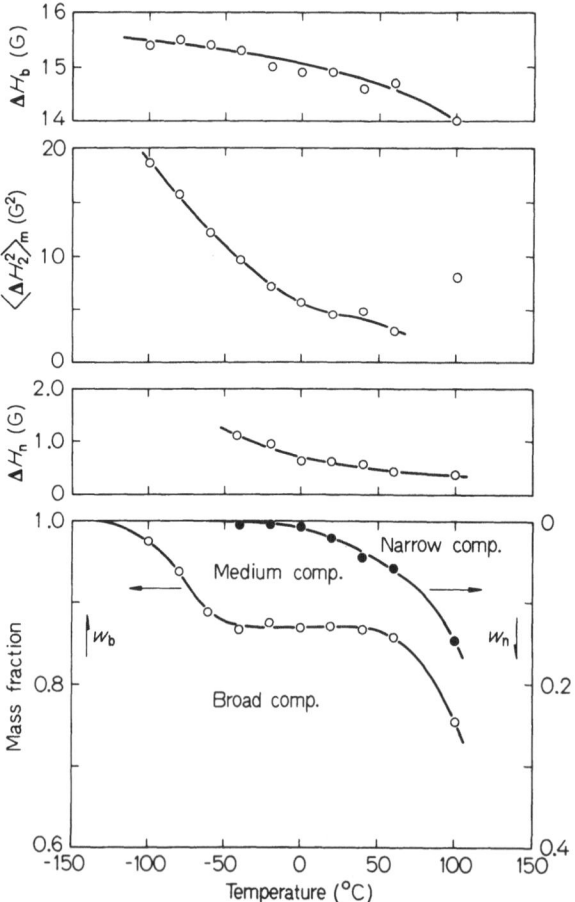

Fig. 12. Mass fractions, line-widths or second moment of three components as a function of temperature for a bulk-crystal with $\bar{M}n$ = 90000 [67)]

In Fig. 12 the mass fractions and the line widths or the second moment of the three components are plotted against temperature. The temperature dependency of the three components observed for this sample is primarily similar to those obtained by Bergmann and Nawotki[13, 15] for unfractionated polyethylene samples. It will be understood as discussed by them in terms of changes of correlation times τ_c for various modes of motions in the structure, corresponding to the γ-, β-, and α-relaxations observed in dielectric and mechanical measurements. At about $-150\,°C$ only the broad component is detected. This shows that all the τ_c associated with various modes of molecular motion in the structure are longer than 10^{-5} sec and the whole structure is considered to be in the rigid state by the NMR measurement. The appearance of the medium component is observed at a temperature higher than $-150\,°C$. This is thought to correspond to the γ-relaxation. At these temperatures some of the τ_c for a local molecular motion such as the hindered rotation of methylene groups about the C–C bonds in the amorphous regions are considered to reach a time less than 10^{-5} sec so as to contribute to the medium component. As the temperature rises, this local molecular motion with $\tau_c < 10^{-5}$ sec is considered to spread over the whole noncrystalline region, resulting in the increase of the w_m and decrease of the w_b.

The narrow component appears above $-40\,°C$ and increases with decreasing medium component. This appearance and increase of the w_n at temperature between $-40°$ and $+50\,°C$ will correspond to the β-relaxation process observed in the mechanical and dielectric measurements. This phenomenon will be understood a resulting from the fact that in this temperature range a micro-Brownian molecular motion starts and spreads over the interzonal region with increasing temperature, while remaining methylene groups of the interfacial region experience only the hindered-rotating type of motion.

At temperatures between $-40°$ and $+50\,°C$ the value of w_b shows a stationary level of 0.870 ± 0.003 which corresponds well with the degree of crystallinity (0.867) obtained from the density measurement at room temperature. It is thought that in this temperature range the molecular chains in the noncrystalline region undergo either hindered-rotational or micro-Brownian motions and those in the crystalline region remain as in the rigid lattice in the NMR measurement. Therefore, in such a temperature range the NMR data can be successfully analyzed in connection with the morphologic phase structure of samples.

As the temperature further rises, the w_b seems to decrease rather abruptly at about $80\,°C$. This phenomenon will correspond to the so-called α-relaxation associated with the onset of some local molecular motion in the crystalline region. Sine the three-component analysis was no longer possible in this temperature range $(70-90\,°C)$, there is no data available for detailed discussions at the present time. If some local molecular motions appear in some part of the crystalline region they should contribute to the medium component, resulting in the decrease of w_b. In such a case, the medium component will be contributed by protons in two regions: the noncrystalline (interfacial) and crystalline regions, in both of which the rotation of methylene groups about the C–C bond is hindered, but to different degrees. Thus the NMR spectrum should be analyzed in terms of contributions from four components: a rigid crystalline region, a crystalline region with local molecular mobility, a noncrystalline region (interfacial region) with limited molecular mobility, and a noncrystalline region (interzonal region) with liquidlike molecular mobility.

However, the three-component analysis becomes possible again at 100 °C, giving reduced w_b and increased w_n as shown in Fig. 12. It is thought that at 100 °C the micro-Brownian molecular motion has spread over the whole of the noncrystalline regions and the medium component due to the interfacial region has disappeared. As a result the structure would come to comprise three regions with different molecular mobilities, so enabling three-component analysis, but in this case the structural unit corresponding to the medium component of the spectrum must be somewhat different. At this temperature the medium component will not be contributed by the noncrystalline region but only by a crystalline region with local molecular mobility. The larger value of $\langle \Delta H_2^2 \rangle_m$ shown in Fig. 12 is well explained by this situation.

Thus the temperature dependency of the three components for this sample with a molecular weight of 90000 has been adequately explained in terms of the α-, β-, and γ-relaxation phenomena associated with the detailed phase structure of the sample.

2. Sample with a Low Molecular Weight

As discussed in the section B of this chapter, the sample with a very low molecular weight is predominantly composed of a lamellar crystalline region, with a minor amount of interfacial region, and no liquidlike interzonal region at room temperature, as can be schematically depicted in either Fig. 10 (B) or (C). The interfacial region comprises relatively short methylene sequences with very limited mobility that are excluded from the crystalline region. This characteristic feature of the phase structure is also reflected in the temperature dependence of the NMR spectrum.

In Figure 13 the mass fractions and the line-widths or second moment for a sample with a molecular weight of 13300 are plotted against temperature. It is seen that the medium component appears at a temperature above −150 °C in an essentially similar manner to samples with higher molecular weights, corresponding to the γ-relaxation due to the onset of a local molecular motion in the noncrystalline region. The w_m increases with decreasing w_b as the temperature rises. But the w_m does not increase so much and the w_b holds a very high stationary level of 0.941 ± 0.009 in a temperature range of −50° ~ + 60 °C, which corresponds well with the degree of crystallinity (0.931) obtained from the density measurement at room temperature. The onset of the narrow component due to the β-relaxation is hardly detected to any appreciable extent until higher temperatures. A significant mass fraction of w_n larger than 0.005 is observed at temperatures only above +50 °C. These temperature behaviours of the mass fractions can be understood by the fact that, since the interfacial region comprises mostly parallel molecular chains excluded from the crystalline region, the allowable conformations of molecular chains are much restricted and molecular motion cannot appear beyond a certain limit. This situation is more intimately reflected in a change in the second moment of the medium component against temperature.

As shown in the figure, the second moment of the medium component $\langle \Delta H_2^2 \rangle_m$ exhibits very characteristic S-type behavior against temperature. It decreases in accordance with the increase of the w_m as the temperature rises, but it holds a level

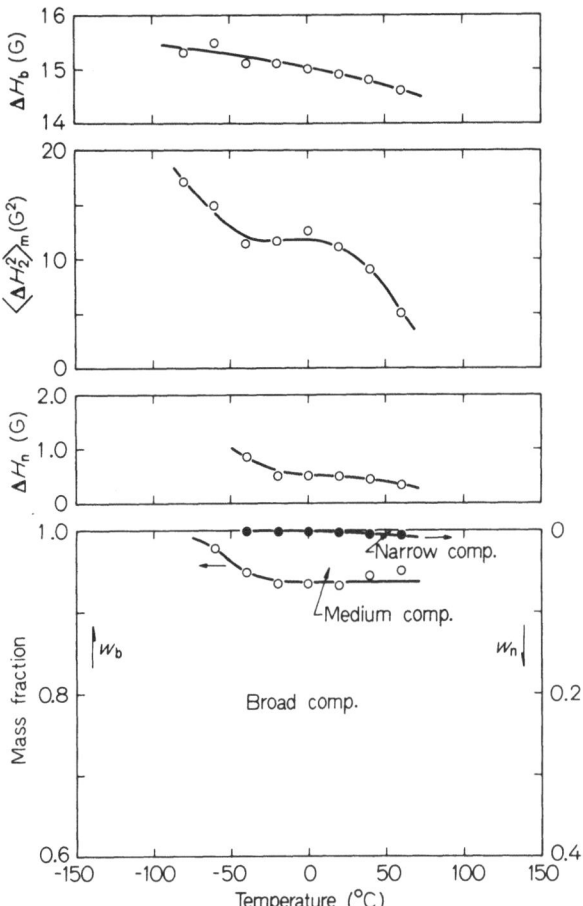

Fig. 13. Mass fractions, line-widths or second moment of three components as a function of temperature for a bulk-crystal with $\bar{M}\eta$ = 13 300 [67]

of about 12 G^2 above room temperature and decreases again at temperatures above 40 °C. This characteristic two-step change in the $\langle \Delta H_2^2 \rangle_m$ is also recognized for stretched samples as will be discussed in a later chapter. Therefore, the molecular chains in the noncrystalline region may have a similar conformation for these two kinds of sample. In any case, such a behavior of $\langle \Delta H_2^2 \rangle_m$ in the vicinity of room temperature is strongly associated with the restricted molecular mobility due to the highly limited conformations of the molecular chains in the noncrystalline region. The delay of onset of the nattow component until higher temperatures will also be caused by this limited molecular mobility.

3. Sample with a High Molecular Weight

As already shown in Fig. 9, if the molecular weight exceeds 400000 the line-widths or second moment of the three components at room temperature nearly reach sta-

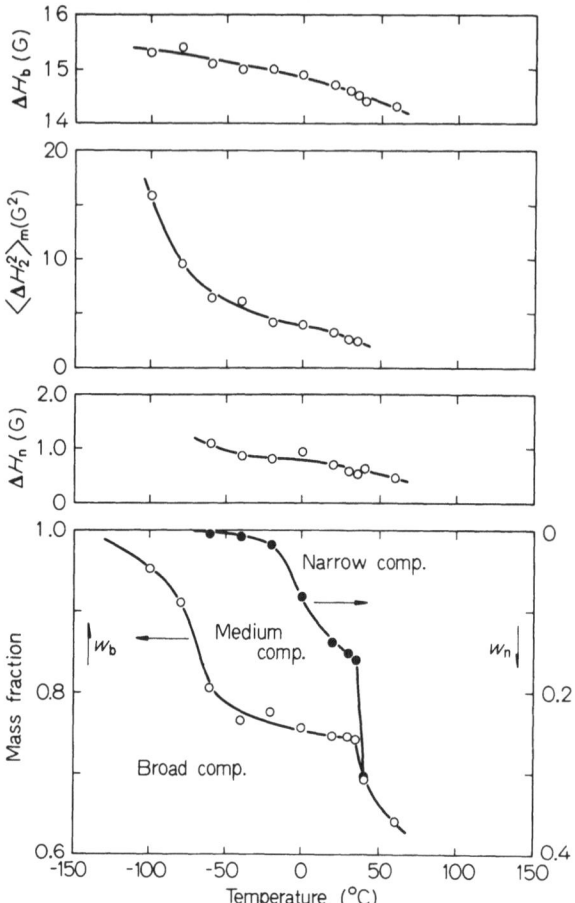

Fig. 14. Mass fractions, line-widths or second moment of three components as a function of temperature for a bulk-crystal with $\bar{M}\eta = 431\,000$ [67)]

tionary levels, although the w_b still decreases with increasing w_m and w_n as the molecular weight further increases. The bulk-crystals with molecular weights in this range should be characterized by a phase structure with a very low degree of crystallinity and larger molecular mobilities in the noncrystalline regions. This characteristic feature in the phase structure is also reflected in the temperature dependency of the three components.

In Fig. 14, the mass fractions and the line-widths or second moment of the three components are plotted against temperature for the sample with a molecular weight of 431 000. In good agreement with the low degree of crystallinity, the medium component appears at a temperature well below $-100\,°C$ and rather rapidly increases with decreasing w_b (γ-relaxation). As expected from the rather versatile conformations of molecular chains as depicted schematically in Fig. 10 (E), the local molecular motion spreads over the whole noncrystalline region, and the mobility increases. The w_m reaches about 0.2 and the $\langle\Delta H_2^2\rangle_m$ decreases to below about 6 G^2 at $-60\,°C$, showing larger molecular motions in the noncrystalline region. In accordance with such smaller values of $\langle \Delta H_2^2 \rangle_m$, the narrow component appears to an appreciable amount at $-50\,°C$ and rapidly increases to a mass fraction of about 0.16 at 30 °C.

As the temperature further rises, this sample shows a very enhanced change in the spectrum at temperatures between 35 and 40 °C. The w_b begins to decrease and the w_m suddenly disappears with increasing w_n. Thus, the spectrum at 60 °C could be well analyzed into two components, the broad and narrow. It may correspond to a two-phase structure composed of the crystalline region and a noncrystalline region with micro-Brownian molecular motion.

It should be noted here that the temperature range (35–40 °C), in which the enhanced change in the spectrum is recognized for this sample, is considerably lower than temperatures associated with α-relaxation (see Fig. 12). Accordingly, the enhanced change in the spectrum cannot be caused by the onset of a local molecular motion in the crystalline region. For samples with larger molecular weights, a partial melting of the crystalline region will occur even at the lower temperatures, while it may not occur appreciably in samples with smaller molecular weights. When samples with large molecular weights are crystallized isothermally at 130 °C from the melt, the mass fraction transformed to the crystalline phase is generally limited, but upon cooling to room temperature an additional amount of mass, *e.g.*, about 0.1, is crystallized[74]. Although the structure of the crystalline content thus crystallized additionally is unknown, it will comprise interlamellar crystallites with smaller dimensions which are rather unstable and associated with lower melting temperatures. Therefore, such small crystallites are thought to melt[f] appreciably at temperatures even lower than 40 °C. If a partial melting of such interlamellar crystallites takes place, even to a small extent, the molecular mobility in the noncrystalline region will be much raised, so that micro-Brownian molecular motion can take place in the whole non-crystalline region, resulting in the disappearance of the medium component.

D. Conclusion

In this chapter we have reviewed the NMR studies on bulk-crystallized samples with different molecular weights over a wide range of temperature. The three-component spectrum analysis gives us very detailed information on the multiphase structure of the samples in terms of the molecular mobility or relaxation modes associated with each phase. It is concluded that the bulk-crystals generally have a structure composed of lamellalike crystallites with interfacial and interzonal materials, associated with limited and liquidlike molecular mobilities, respectively. The relative content of these components as well as the molecular conformation or mobility in each component vary over a very wide range, depending strongly on the molecular weight.

Samples of very low molecular weight are predominantly composed of a lamellar crystalline region with a minor amount of the interfacial region but no liquidlike interzonal region. As the molecular weight increases beyond 40000, an interzonal

---

[f]   In fact, the decreasing of the degree of crystallinity with elevating temperature is evident in the dilatometric data of Fatou and Mandelkern[74] even at such low temperatures for samples with larger molecular weights.

region with a liquidlike character associated with high molecular mobility is produced. Above a molecular weight of 100000, this liquidlike character becomes pronounced, with an increase in molecular mobility in the interfacial region. This detailed change of the multiphase structure of samples is also reflected in the temperature dependency of the phase structure in connection with different relaxation processes recognized in other measurements as α-, β-, and γ-relaxations.

V. Samples Crystallized from Dilute Solution

A. Introduction

It is well known from electron-microscopic observations[4−6] that linear polyethylene, when crystallized from dilute solutions, forms generally lozenge-shaped platelets or lamellalike crystallites. The lateral dimensions of such crystallites are usually of the order of several microns, while they are about 100 Å thick. Furthermore, it is evidenced by selected area electron diffraction studies[5, 6] that the molecular chain axes are oriented perpendicular to the wide faces of the lamella. Since the extended molecular chain length is much larger than the lamellar thickness for samples with usual molecular weights, it is necessary that a given molecule participates in a crystallite repeatedly in order to satisfy the requirements of chain length, crystallite thickness, and chain orientation. Therefore, the noncrystalline region, the existence of which has been clearly evidenced also for samples crystallized from dilute solution (abbreviated hereafter as solution-grown samples or solution-crystals) by their lower density and enthalpy change in fusion compared to the perfect crystalline material, may have a unique conformation of molecular chains. Such characteristics in molecular chains are also very much reflected in the NMR spectrum[67]. In this chapter we discuss the phase structure of the samples by analyzing the spectrum over a wide range of temperatures.

B. Spectrum Analysis at Room Temperature

The spectrum for samples with a very low molecular weight, i.e., lower than about 1000, is fairly independent of the mode of crystallization, whether from the melt or from dilute solution. The spectrum is characterized by a very large broad component ($w_b \simeq 0.95$) and a medium component with a large second moment, but no narrow component. In such samples the extended molecular chain length will be comparable to or slightly larger than the lamella thickness. The conformation of molecular chains to form the lamellar crystallites will be similar to that depicted schematically in Fig. 10 (B), independent of the crystallization mode.

However, samples with molecular weights higher than several thousands, if crystallized from dilute solution, exhibit a unique type of spectrum, fairly independent of the detailed conditions of the crystallization such as solvent or temperature. Figure 15 shows the spectra at room temperature for three samples with different molecular weights that were crystallized from 0.08 toluene solution isothermally at 85 °C. The molecular weights vary over a very wide range but the spectra are all similar. The composite curve of the three components coincides closely with the observed spectrum, except for some minor deviation[g] in larger field intensities, demonstrating the adequacy of the three-component analysis for solution-crystals also.

The numerical data of the analysis are summarized in Table 1. Although the molecular weights differ by a factor of about 200, the results are rather similar.

Table 1. Three-component analysis of the spectrum at 20 °C for solution-grown samples with different molecular weights

Sample	$\bar{M}\eta$	Mass fraction			Line-width (G)			Sec. mom.(G^2)
		w_b	w_m	w_n	ΔH_b	ΔH_m	ΔH_n	$\langle \Delta H_2^2 \rangle_m$
SS 10-5	17600	0.861	0.136	0.003	14.7[a]	5.5	0.81	8.4
SS 3-2	106000	0.825	0.169	0.006	14.8	5.4	1.04	8.4
SH 1-1	3400000	0.835	0.160	0.005	14.8	5.5	1.00	8.4

[a] ΔH_b of 14.7 G corresponds to a second moment $\langle \Delta H_2^2 \rangle_b$ of 21.9 G^2.

This is quite contrary to the case for the bulk-crystals, where, as reviewed already, the result depends very much on the molecular weight. This clearly suggests a unique phase structure independent of molecular weight for solution-crystals, which is in good agreement with other measurements of the thermodynamic quantities of the samples. Independent of the molecular weight, it is noted that the w_m is as large as 0.14–0.17 while the second moment is rather large at 8.4 G^2. The narrow component is scanty ($w_n \leq 0.006$), but the line-width is significantly large at 0.8–1.0 G. The mass fraction of the broad component is of the order of 0.83–0.86, in good accord with the degree of crystallinity evaluated from the macroscopic density, and the line-width is about 14.8 G which corresponds to the values for bulk-crystals with larger molecular weights. These results not only indicate a highly ordered state of the crystalline region but also clearly suggest the existence of a substantially amorphous overlayer with less molecular mobility in the lamellar structure.

The spectrum analysis shows that the solution crystals comprise about 85% lamellalike crystalline region and about 15% interlamellar amorphous overlayer, with a very minor liquidlike amorphous component at room temperature. However, the structure of the amorphous overlayer seems to be somewhat different from that in the bulk-crystals. For the bulk-crystals the second moment of the medium component ascribed to the amorphous overlayer appreciably decreases with increasing molecular weight. The quantity for the solution crystals, however, stays unaltered at a value as high as 8.4 G^2 even if the molecular weight becomes as large as 3.4×10^6. The

[g] Similar deviation was also recognized by Bergmann[14] for solution-grown samples. This will be caused by the elementary spectrum used for the broad component. Detailed discussion is outside the scope of this review.

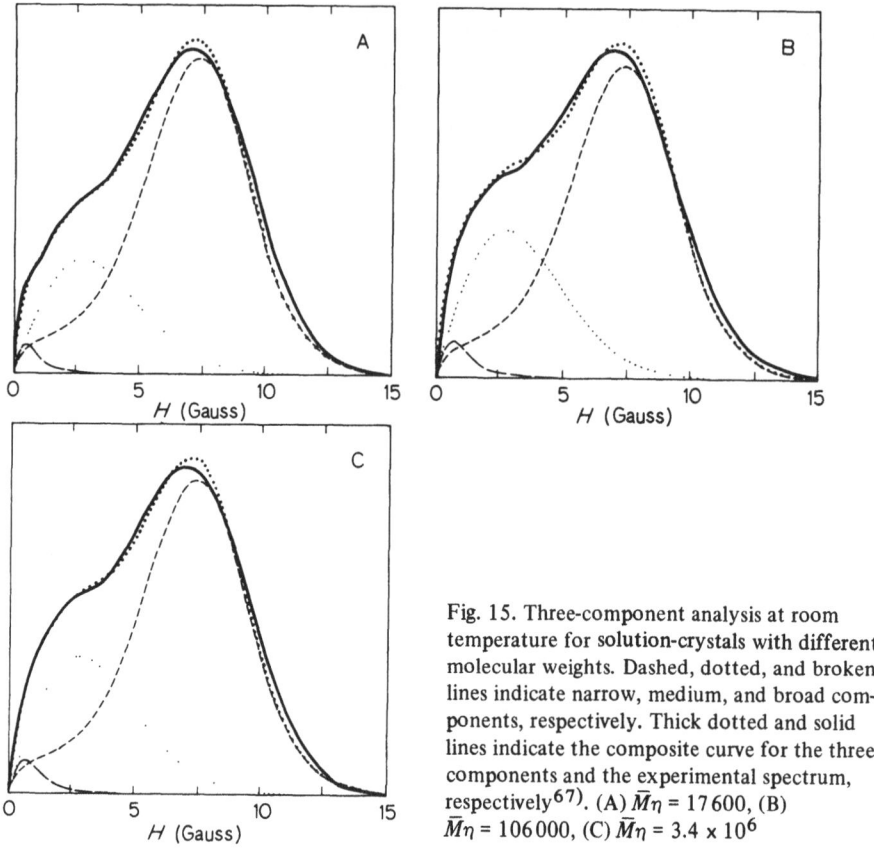

Fig. 15. Three-component analysis at room temperature for solution-crystals with different molecular weights. Dashed, dotted, and broken lines indicate narrow, medium, and broad components, respectively. Thick dotted and solid lines indicate the composite curve for the three components and the experimental spectrum, respectively[67]. (A) $\bar{M}\eta = 17\,600$, (B) $\bar{M}\eta = 106\,000$, (C) $\bar{M}\eta = 3.4 \times 10^6$

value of 8.4 G^2 seems to be appreciably smaller than the value expected according to the so-called regularly folded molecular chain model, where the molecular rotation about the C—C bond is completely inhibited. Nevertheless, the rather high level of the second moment independent of the molecular weight must reflect such a unique conformation of molecular chains in the overlayer that molecular motion is strongly restricted by the existence of crystallites. We will discuss this matter after seeing the temperature dependence of the spectrum analysis.

C. Temperature Dependence of the Phase Structure

We have found by examining the spectrum for the solution-grown samples at room temperature that their phase structure is composed of lamellalike crystallites and an amorphous overlayer having limited molecular mobility, with a very small amount of liquidlike amorphous content. This unique phase structure is reflected in more detail in the temperature dependence of the spectrum.

Figure 16 shows the three-component analysis of spectra for a sample with a molecular weight of 106 000 at various temperatures from −150° to +60 °C. The agreement of the composite curves with the observed curves is excellent except for

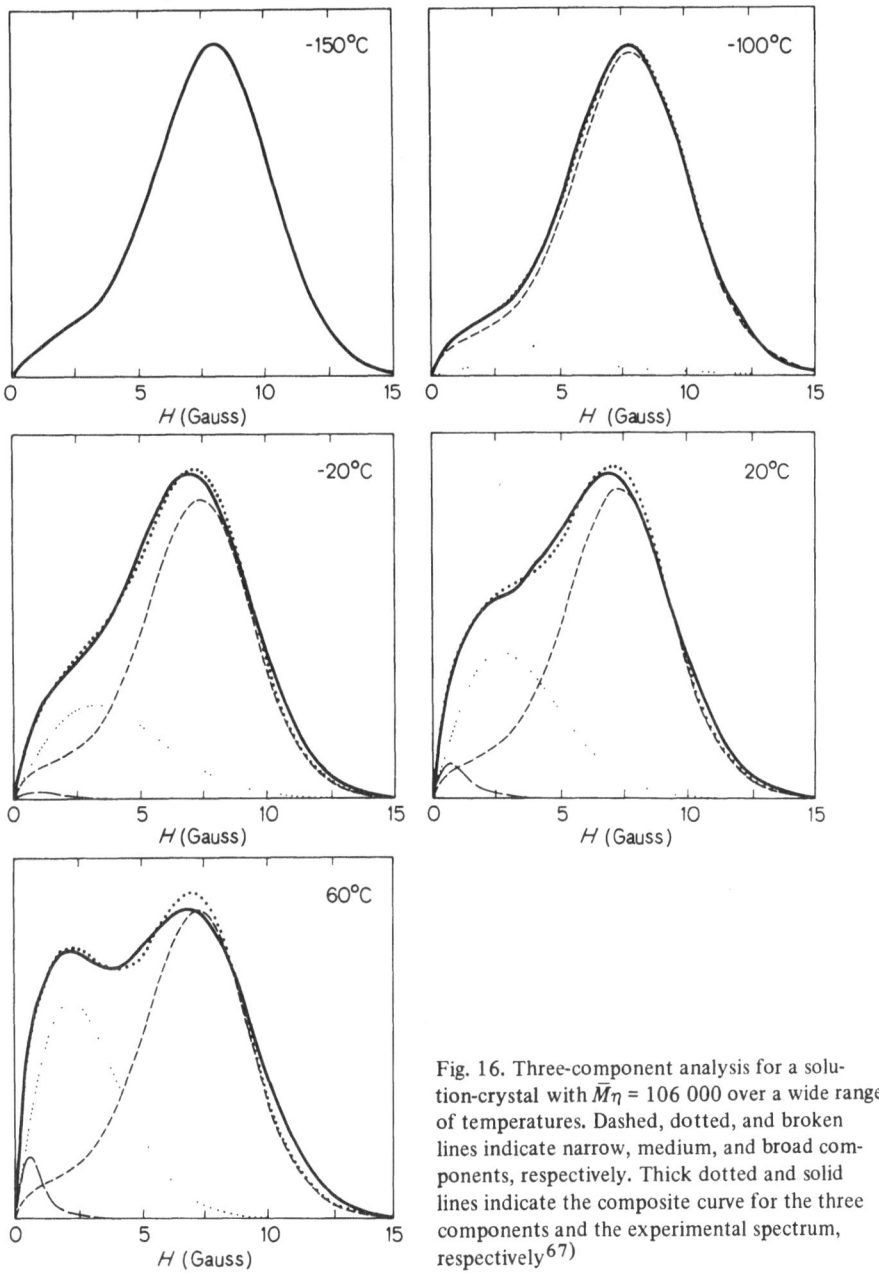

Fig. 16. Three-component analysis for a solution-crystal with $\overline{M}\eta$ = 106 000 over a wide range of temperatures. Dashed, dotted, and broken lines indicate narrow, medium, and broad components, respectively. Thick dotted and solid lines indicate the composite curve for the three components and the experimental spectrum, respectively[67]

minor deviation in the higher field intensities. In Fig. 17 are plotted the mass fractions and the line-widths or second moment of the three components against temperature, for two samples with molecular weights of 106000 and 3400000. Although the molecular weights differ by a factor of more than 30, data for the two samples make families of curves, indicating that the solution-grown samples have a unique phase structure independent of molecular weight. It is seen that the medium com-

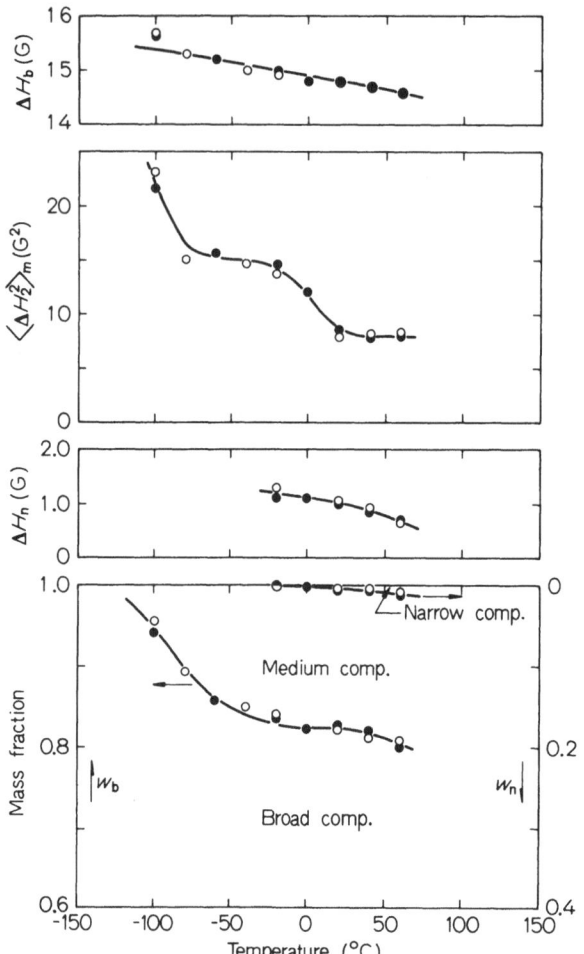

Fig. 17. Mass fractions, line-widths or second moment of three components as a function of temperature for solution-crystals with different molecular weights[67]. o: $\bar{M}\eta = 106\,000$, •: $\bar{M}\eta = 3.4 \times 10^6$

ponent appears at a temperature above $-150\,°C$, corresponding to the γ-relaxation due to the onset of local molecular motion in the noncrystalline content. The narrow component seems to appear at about $-20\,°C$ but does not increase appreciably. Its level is still less than 0.007 even at $60\,°C$. The $\langle\Delta H_2^2\rangle_m$ decreases with increasing w_m but it holds a level above $14\,G^2$ up to $-20\,°C$. It decreases as the narrow component, if any, appears, but holds as high a level as $8\,G^2$ up to $60\,°C$.

These results of the spectrum analysis are quite different from those for the bulk-crystallized samples. In the bulk-crystals with larger molecular weights, after the w_m goes through a maximum it decreases with increasing w_n as the temperature rises. This was considered to be the result of micro-Brownian molecular motion spreading through a major part of the noncrystalline region with increasing temperature. For the solution-crystals, however, such a decrease in w_m is not found. Not only does the w_m stay at a high level about 0.18, but also the $\langle \Delta H_2^2 \rangle_m$ holds a rather high level up to about 60 °C, above which the spectrum analysis could not be made because of the onset of an α-relaxation associated with the crystalline region.

The apparent difference recognized here in the spectrum analysis between the two types of sample should be considered in terms of molecular chain conformation in the interlamellar amorphous contents. Both types of samples, except where molecular weight is very low, have similar crystalline structures, comprising lamellar crystallites, of which the lamellar thickness is usually much smaller than the extended molecular chain length. Therefore, it is necessary that a given molecular chain participates repeatedly in lamellar crystalline regions, penetrating perpendicular to the lamella face. The noncrystalline region is composed of molecular chains, each of which emerges from a lamellar crystallite and returns to the same or another crystallite. In the bulk-crystals a molecular chain which has emerged from a crystallite does not necessarily return to the same crystallite, but can enter another crystallite after 'swimming' in the noncrystalline region, which exhibits a liquidlike character, as schematically depicted in Fig. 10 (D). The conformation of such a molecular chain in the noncrystalline region should be rather versatile. Hence, with increasing temperature such molecular chains exhibit not only a local motion but also a micro-Brownian motion. Thus the increase of the w_n and concomitant decrease of the w_m with increasing temperature can be adequately understood.

On the other hand, the conformation of molecular chains in the noncrystalline region of solution-crystals is thought to be a little different. When crystallized from dilute solution, the number of molecular chains available for the growth of a crystallite in its incipient stage is limited, owing to the low concentration of the material. The growth of a crystallite should take place mostly by an intramolecular aggregation or crystallization. Hence, a molecular chain which has emerged from a lamellar crystallite must return mostly to the same crystallite. Therefore, the conformation of molecular chains in the noncrystalline region is thought to be less versatile and their length is thought to be relatively short. Such molecular chains will exhibit a local molecular motion but they cannot exhibit larger molecular motion such as micro-Brownian motion even at high temperatures. The unique temperature dependency of solution-crystals shown in the spectrum analysis will reflect this unique conformation of molecular chains in the noncrystalline interlamellar material.

D. Conclusion

NMR spectroscopy over a wide range of temperatures shows that the solution-grown samples have a unique phase structure independent of the molecular weight. It is revealed that their structure is composed of lamellar crystallites, with a noncrystalline interfacial overlayer as large as 15%. Molecular motion in the interfacial overlayer is much limited, and the noncrystalline component with liquidlike mobility such as was recognized generally for the bulk-crystals could not appear appreciably even at higher temperatures. Such limited molecular mobility could well be explained in terms of a rather restricted conformation of the molecular chains in the noncrystalline region, considering the special mode of crystallization from dilute solution.

VI. Fibers

A. Introduction

Crystalline polymers are generally transformed into the so-called fiber structure by uniaxial stretching at temperatures between the glass and melting temperatures of the polymers. In such fiber structures X-ray diffraction and optical birefringence analyses reveal that molecular chains in both the crystalline and the amorphous regions are preferentially oriented parallel to the stretching direction. However, their detailed phase structure is still unknown, although it is of great importance in relation to the tensile properties in practical use. This chapter reviews the phase structure of linear polyethylene fibers drawn to different draw ratios as revealed by the spectrum over a wide range of temperatures[75].

Filaments about 0.12 mm in diameter were obtained from unfractionated high-density polyethylene ($\bar{M}\eta = 8.0 \times 10^4$) by a melt-spinning at 220 °C with a draft ratio of about 17. They were next stretched at 100 °C in polyethylene glycol with a molecular weight of 380—400 to different draw ratios, which are defined as the ratios of the drawn length to the original. NMR spectroscopy was carried out for the drawn filaments randomly packed into a glass tube 18 mm diameter. The three-component analysis of the spectrum at different temperatures was performed; the results are discussed in relation to the phase structure of samples.

B. Spectrum Analysis at Room Temperature

Figure 18 shows the spectra for samples at draw ratios of 1, 2, 4, 6, 8, and 10, with the three-component analyses. It can be seen that all the spectra give a good analysis using the three components, although they differ in shape to a great extent, depending on the draw ratio. In the spectra of the samples with higher draw ratios, three distinct peaks can be recognized, as reported for the stretched samples[16, 76]. The numerical results are summarized in Table 2.

The orientation factors f_c and f_a, which characterize the orientation of molecular chains in the crystalline and amorphous regions respectively, with respect to the stretching direction, were obtained from X-ray diffraction and birefringence measurements together with the measured density[77, 78]. As is clearly seen from the table, both f_c and f_a increase with increasing draw ratio. This suggests that a typical fiber structure is produced by the drawing procedure. The molecular chains, not only in the crystalline region but also in the noncrystalline region, are preferentially oriented parallel to the stretching direction with increasing draw ratio.

It is to be noted here that the mass fraction of the broad component is not altered appreciably by drawing, so that $w_b \simeq 0.689 \pm 0.009$ independent of the draw ratio. On the other hand, the density increases from 0.9502 to 0.9662 with increasing draw ratio. If a constant density is assumed for the amorphous material, this result must show the increase of the degree of crystallinity with stretching. However, the density of the amorphous material might be changed appreciably with stretching. For example, it was reported by Peterlin and others[79] using X-ray scattering analysis

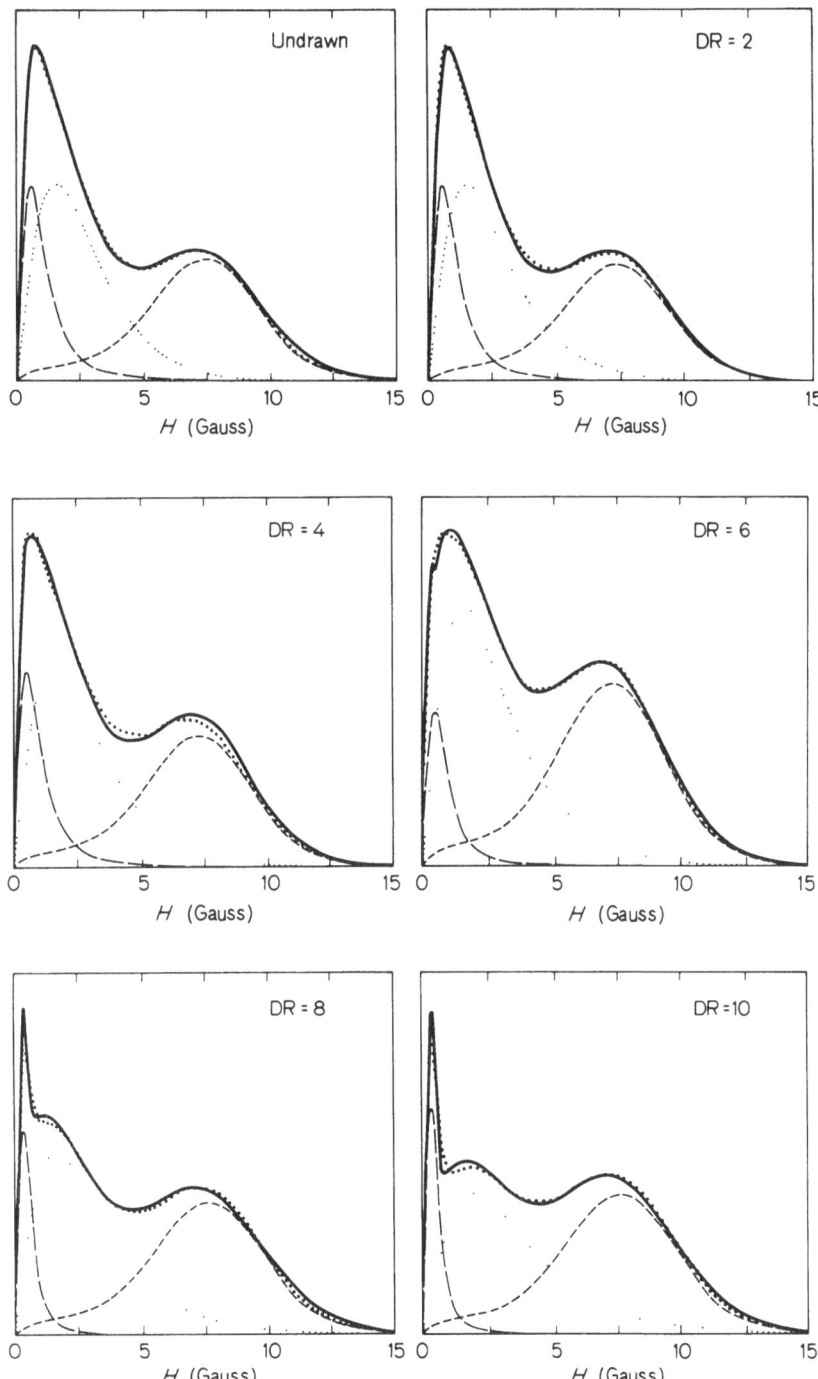

Fig. 18. Three-component analysis at room temperature for fibers drawn to different degrees. The draw ratio (DR) is indicated in each figure. Dashed, dotted, and broken lines indicate narrow, medium, and broad components, respectively. Thick dotted and solid lines indicate the composite curve for the three components and the experimental spectrum, respectively[75]

Table 2. Three-component analysis for drawn filaments

Draw ratio	Density at 30 °C (g/cm³)	Orientation factor[a]		Mass fraction			Line-width or sec. mom. in G or G², resp.			
		f_c	f_a	w_b	w_m	w_n	ΔH_b	ΔH_{mb}	$\langle \Delta H^2 \rangle_m$	ΔH_n
1	0.9502	0.383	0.293	0.685	0.257	0.058	15.0	3.19	5.9	1.00
2	0.9512	0.615	0.494	0.679	0.263	0.058	14.8	3.06	6.3	0.98
4	0.9532	0.739	0.733	0.680	0.268	0.052	14.8	3.25	6.9	0.98
6	0.9546	0.855	0.835	0.691	0.288	0.021	14.9	3.28	8.4	0.80
8	0.9602	0.924	0.884	0.698	0.287	0.015	15.2	3.36	9.5	0.45
10	0.9662	0.972	0.898	0.698	0.289	0.013	15.2	4.06	11.3	0.40

a) The orientation factor is defined as $f = (3 \langle \cos^2 \varphi \rangle - 1)/2$, where φ is the angle of a molecular chain to the stretching direction.

that the density of the amorphous content was increased when stretched to high extents. If the unit-cell density is assumed to be 1.00 and the density of the amorphous region is assumed to change from 0.85 to 0.89 with drawing referring to their result, the degree of crystallinity is estimated to be unchanged within experimental error at 0.698 ± 0.04. This value is in good accord with the w_b obtained by the NMR analysis. If different values are assumed for the density of the amorphous region, the result may be changed appreciably. However, at the present time we may conclude that the degree of crystallinity is not changed appreciably by drawing and the w_b corresponds well with the crystalline fraction in undeformed samples.

On the other hand, the analysis results in respect of the noncrystalline components are greatly changed with drawing. A significant increase in $\langle \Delta H_2^2 \rangle_m$ is recognized with increasing draw ratio. This is considered as a result of the molecular chains in the noncrystalline region being oriented, as is shown by the increase of f_a, and the versatility of the conformation is confined upon stretching.

Furthermore, a significant decrease in both w_n and ΔH_n is recognized with increasing draw ratio. It shows that some amorphous molecular chains with liquidlike mobility are stretched with macroscopic drawing and come to contribute to the medium component. Some narrow component remains even after drawing to the highest draw ratio. This will be contributed from protons that belong to methyl endgroups or methylene groups adjacent to them. Since those protons are thought to be insensible to the macroscopic stretching and to be appreciably mobile, the decrease of ΔH_n with increasing draw ratio will be well understood.

In conclusion, the spectrum analysis at room temperature shows several important changes in the phase structure upon stretching at 100 °C. Before stretching, the sample filament has a phase structure approximately similar to the bulk-crystals discussed in the foregoing chapter, although some preferential orientation of molecular chains parallel to the filament direction exists, as suggested by the f_c and f_a larger than zero. The phase structure comprises lamellalike crystallites and amorphous interfacial and interzonal materials, and it is drastically changed with macroscopic stretching at high temperature, and the so-called fiber structure is produced.

In the fiber structure, molecular chains in the crystalline region are highly oriented parallel to the fiber axis and a majority of molecular chains in the noncrystalline region are also highly oriented, so that both the conformational versatility and the mobility of molecular chains are much limited. But even in highly drawn fibers a significant, albeit minor, amount of amorphous material with a liquidlike mobility still exists.

C. Temperature Dependence of the Phase Structure

We have reviewed the phase structure of the fiber samples as a function of the draw ratio in terms of molecular mobility by analyzing the spectrum at room temperature. In this section we consider the phase structure over a wide range of temperatures.

In Fig. 19 the mass fractions and the line-widths or the second moment of the three components for samples with draw ratios of 1, 6, and 10 are plotted against temperature. It is clearly seen that the temperature dependence of the three com-

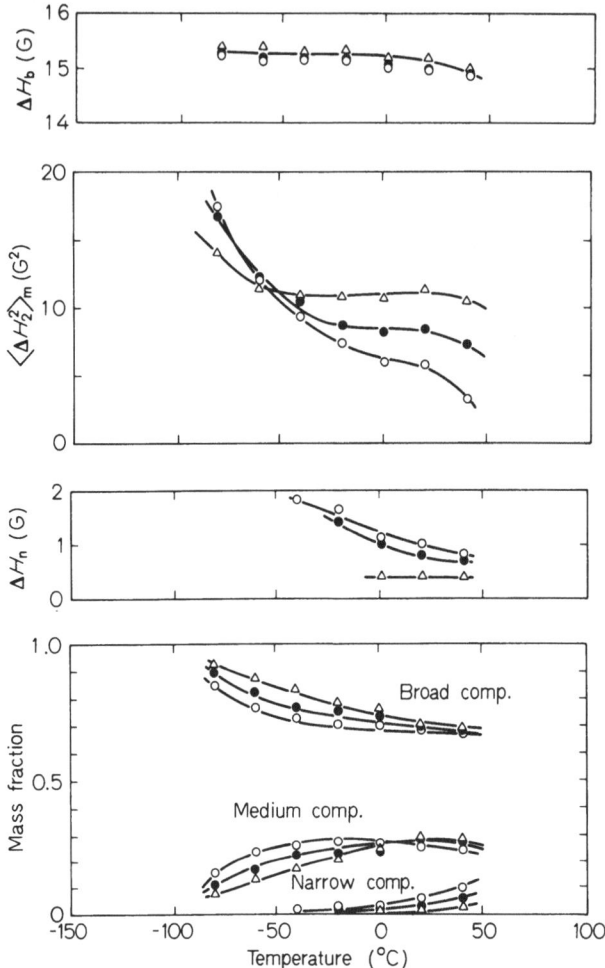

Fig. 19. Mass fractions, line-widths or second moment of three components as a function of temperature for fibers drawn to different degrees [75]. o: undrawn, ●: drawn 6-fold, △: drawn 10-fold

ponents can also be essentially understood in terms of the β- and γ-relaxation phenomena, as was shown for the bulk- and solution-crystallized samples. However, the detailed temperature dependency seems to be greatly changed with the draw ratio. The medium component starts to appear at a higher temperature and increases more gradually as the draw ratio increases. Concomitantly the mass fraction of the broad component decreases more gradually and reaches and asymptotic level at a higher temperature. This suggests that the γ-relaxation temperature tends to shift to a higher temperature with increasing draw ratio. As a result, the temperature region in which the w_b actually corresponds to the crystalline fraction is shifted to a higher temperature region, above room temperature, for the highly drawn samples, while the w_b does represent the crystalline mass fraction in a temperature range of -40 to

+50 °C for the bulk-crystallized sample, as shown in Fig. 12. A similar effect of drawing is also evident in the behavior of the narrow component, suggesting that the β-relaxation process is shifted to a higher temperature with increasing draw ratio.

Such increases of γ- and β-relaxation temperatures with increasing draw ratio are thought to be due to the limited mobility of molecular chains in the noncrystalline region. As pointed out in the foregoing section, the amorphous molecular chains align fairly well parallel to the drawing direction. Therefore, the conformational versatility and mobility of such molecular chains are considered to be much restricted.

Such highly restricted molecular mobility is also suggested by a very characteristic change of the $\langle \Delta H_2^2 \rangle_m$ against temperature, as shown Fig. 19. The $\langle \Delta H_2^2 \rangle_m$ of the 10-fold drawn sample decreases in accordance with the increase of the w_m as the temperature rises and holds a very high level above 10 G^2 in the vicinity of room temperature and decreases again above 40 °C. This characteristic S-type behavior and the high value of $\langle \Delta H_2^2 \rangle_m$ are approximately equivalent to those for the bulk-crystals with a low molecular weight (see Fig. 13). In the bulk-crystals the molecular chains that contribute to the medium component must be composed of short molecular chains that are excluded from the basal plane of lamellar crystallites. Such molecular chains must be in a rather extended chain conformation and comprise a parallel intermolecular alignment perpendicular to the basal plane of a crystallite, as depicted schematically in Fig. 10 (B). The large values and characteristic temperature dependency of $\langle \Delta H_2^2 \rangle_m$ recognized for both kinds of sample may be associated with the highly extended molecular chain conformation in the amorphous region.

For the 10-fold drawn sample the ΔH_n remains at a very low level, about 0.4 G, over the temperature range examined. As discussed already, the narrow component for such a highly drawn sample will be contributed mainly by protons belonging to methyl end-groups or adjacent methylene groups insensible to the macroscopic drawing. Such components will be fairly mobile even at lower temperatures. The very small value of ΔH_n for the highly drawn sample can thus be easily understood.

D. Conclusion

NMR spectrum analysis shows that the phase structure of the drawn fibers is quite different from that of undeformed samples. The undrawn filament obtained by a melt-spinning has a phase structure approximately similar to that of bulk-crystals with corresponding molecular weight. However, the structure is drastically changed with drawing. The highly drawn filaments comprise crystalline material as well as a noncrystalline component of more than 25%, which is similar to undeformed bulk-crystals. Nevertheless, the molecular mobility and conformation of molecular chains in the noncrystalline component are quite different. The majority of molecular chains are almost fully extended parallel to the fiber axis and are associated with a very limited molecular mobility over a wide range of temperatures. However, noncrystalline. material with liquidlike mobility is significantly present, small though the amount may be, even in the 10-fold drawn sample. This conclusion has been further supported by spectroscopy[75] for samples immersed in such a nonprotonated solvent as CCl_4, which enhances molecular mobility preferentially in the amorphous region.

VII. Concluding Remarks

In this article we have reviewed our recent work with NMR analysis on various kinds of linear polyethylene samples. It has become evident that the refined NMR analysis gives us much important information on the phase structure of samples in terms of molecular mobility, and establishes that there is no unified phase structure for polymer samples. The phase structure of samples varies over a very wide range, depending strongly on the sort of samples involved as well as on the mode of crystallization or the history of those samples. We should emphasize that there are significant differences in phase structure among the bulk-crystals, the solution-crystals, and the fiber samples, particularly in the conformation of molecular chains in the noncrystalline content. We should not confuse these phase structures with each other. The phase structures are evidently different, sample by sample, as their macroscopic properties also differ one from another.

At this end, we should make a note in relation to the macroscopically lozenge-shaped lamellalike crystallites for the solution-grown samples. For such lamellar structure, it has been proposed[4-6] that a molecular chain participates repeatedly in a lamellar crystallite by folding regularly with adjacent reentry. However, this crystal model evidently fails in explanation for the existence of large amounts of noncrystalline content and various noncrystalline properties. Alternatively, it has been proposed[80] that adjacent reentry does not occur and loops with random lengths connect the crystalline sequences more or less at random. This model permits a large disordered overlayer which is not allowed in the former model. The present NMR work is well reconcilable with this latter crystal model.

Doubtless, thermodynamically most stable form will be such one as depicted schematically in Fig. 10 as unpeeled crystal (so-called extended chain crystal), independent of the mode of crystallization. Such a stable crystal is attainable for samples with very low molecular weight. Nevertheless, for samples with higher molecular weights, the lamellar structure with thicknesses much smaller than the extended chain length is only attainable under usual conditions, even if crystallized from either the melt or dilute solution. The reason for the formation of such lamellar structure should not be sought in a very peculiar habit of long chain molecules to tightly fold back but it should be sought in an anisotropic rate of the crystal growth, characteristic of long-chain molecules. Because of the coiled and entangled state of molecular chains prior to the crystallization, it is thought that after a crystal center is nucleated it hardly grows in the direction parallel to the chain axis beyond a certain limit but major growth will be achieved in the direction perpendicular to the chain axis.

Acknowledgement. Much of the work reviewed in this article was achieved with the help of Mr. S.-H. Hyon. We thank him for his enthusiastic cooperation. We also wish to express our sincere thanks to Professor L. Mandelkern of the Florida State University and Professor R. Chūjō of Tokyo Institute of Technology for their useful discussions and comments.

VIII. References

1) Mandelkern, L.: Crystallization of polymers. New York: McGraw-Hill 1964
2) Takayanagi, M.: Pure and Appl. Chem. 15, 555 (1967)
3) Illers, K. H.: Kolloid-Z. Z. Polym. 231, 622 (1969)
4) Till, P. H.: J. Polym. Sci. 24, 301 (1957)
5) Keller, A.: Phil. Mag. 2, 1171 (1957)
6) Fischer, E. W.: Z. Naturforsch. 12a, 753 (1957)
7) Kubo, R., Tomita, K.: J. Phys. Soc. Japan 9, 888 (1954)
8) Odajima, A., Sauer, J. A., Woodward, A. E.: J. Phys. Chem. 66, 718 (1962)
9) Wilson III, C. W., Pake, G. E.: J. Polym. Sci. 10, 503 (1953)
10) Fischer, E. W., Peterlin, A.: Makromol. Chem. 74, 1 (1964)
11) Olf, H. G., Peterlin, A.: Kolloid-Z. Z. Polym. 215, 97 (1967)
12) Bergmann, K., Nawotki, K.: Kolloid-Z. Z. Polym. 219, 132 (1967)
13) Bergmann, K.: Ber. Bunsenges. Phys. Chem. 74, 912 (1970)
14) Bergmann, K., Nawotki, K.: Kolloid-Z. Z. Polym. 250, 1094 (1972)
15) Bergmann, K.: Kolloid-Z. Z. Polym. 251, 962 (1973)
16) Loboda-Čačković, J., Hosemann, R., Wilke, W.: Kolloid-Z. Z. Polym. 235, 1253 (1969)
17) Phaovibul, O., Loboda-Čačković, J., Hosemann, R.: Makromol. Chem. 175, 2991 (1974)
18) Eichhoff, U., Zachmann, H. G.: Ber. Bunsenges. Phys. Chem. 74, 919 (1970)
19) Zachmann, H. G.: J. Polym. Sci. Symposium 43, 111 (1973)
20) Zachmann, H. G.: Kolloid-Z. Z. Polym. 251, 951 (1973)
21) Smith, J. B., Manuel, A. J., Ward, I. M.: Polymer 16, 57 (1975)
22) Fischer, E. W., Goddar, H., Piesczek, W.: J. Polym. Sci. C32, 149 (1971)
23) For example, Allen, G., Peterie, S. E., Ed.: Physical structure of the amorphous state. J. Macromol. Sci. Phys. B12, No. 1 and 2, 1976
24) Kirste, R. G., Kruse, W. A., Schelten, J.: Makromol. Chem. 162, 299 (1972)
25) Schelten, J., Kruse, W. A., Kirste, R. G.: Kolloid-Z. Z. Polym. 251, 919 (1973)
26) Kirste, R. G., Kruse, W. A., Ibel, K.: Polymer 6, 120 (1975)
27) Cotton, J. P., Farnoux, B., Jannink, G., Mons, J., Picot, C.: C. R. Acad. Sci. C275, 175 (1972)
28) Ballard, D. G. H., Wignall, G. D., Schelten, J.: Eur. Polym. J. 9, 965 (1973)
29) Wignall, G. D., Ballard, D. G. H., Schelten, J.: Eur. Polym. J. 10, 861 (1974)
30) Cotton, J. P., Decker, D., Benoit, H., Farnoux, B., Higgins, J., Jannink, G., Ober, R., Picot, C., Des Cloizeaux, J.: Macromolecules 7, 863 (1974)
31) Daoud, M., Cotton, J. P., Farnoux, B., Jannink, F., Sarma, G., Benoit, H., Duplessix, R., Picot, C., De Gennes, P. G.: Macromolecules 8, 804 (1975)
32) Schelten, J., Wignall, G. D., Ballard, D. G. H., Schmatz, W.: Colloid Polym. Sci. 252, 749 (1974)
33) Lieser, G., Fischer, E. W., Ibel, K.: J. Polym. Sci. Polym. Lett. Ed. 13, 39 (1975)
34) Wendorff, J. H., Fischer, E. W.: Kolloid-Z. Z. Polym. 251, 876, 884 (1973)
35) Renninger, A. A., Wicks, G. G., Uhlmann, D. R.: J. Polym. Sci. Polym. Phys. Ed. 13, 1247, 1481 (1975)
36) Reinninger, A. A., Uhlmann, D. R.: J. Polym. Sci. Polym. Phys. Ed. 14, 353, 415 (1976)
37) Hayashi, H., Hamada, F., Nakajima, A.: Macromolecules 9, 543 (1976)
38) Patterson, G. D.: J. Macromol. Sci. Phys. B12, 61 (1976)
39) Flory, P. J.: Principles of polymer chemistry. Ithaca, New York: Cornell Univ. 1953
40) Yeh, G. S. Y., Geil, P. H.: J. Macromol. Sci. B1, 235 (1967)
41) Yeh, G. S. Y.: J. Macromol. Sci. B6, 451, 465 (1972)
42) Gölz, W. L. F., Zachmann, H. G.: Kolloid-Z. Z. Polym. 247, 814 (1971)
43) Horii, F., Kitamaru, R., Suzuki, T.: J. Polym. Sci. Polym. Lett. Ed. 15, 65 (1977)
44) Zachmann, H. Z., Gölz, W.: J. Polym. Sci. Symposium 42, 693 (1973)
45) Gölz, W. L. F., Zachmann, H. G.: Makromol. Chem. 176, 2721 (1975)
46) Miyake, A.: J. Polym. Sci. 28, 476 (1958)
47) Chûjô, R.: J. Phys. Soc. Japan 18, 124 (1963)

48) Russel, A. M., Torchia, D. A.: Rev. Sci. Instr. *33*, 442 (1962)
49) Wilson, G. V. H.: J. Appl. Phys. *34*, 3276 (1963)
50) Slonim, I. Ya., Lyubimov, A. N., Urman, Ya. G., Konovalov, A. G., Varenik, A. F.: Vyso-komolek. Soed. *7*, 245 (1965)
51) Slonim, I. Ya., Lyubimov, A. N.: The NMR of polymers. Translated from Russian Ed. New York: Plenum 1970, p. 155
52) Gutowsky, H. S., Pake, G. E.: J. Chem. Phys. *18*, 162 (1950)
53) Pechhold, W.: Kolloid-Z. Z.Polym. *228*, 1 (1968)
54) Wahlquist, H.: J. Chem. Phys. *35*, 1708 (1961)
55) Smith, G. W.: J. Appl. Phys. *35*, 1217 (1964)
56) Arndt, R.: J. Appl. Phys. *36*, 2522 (1965)
57) Buckmaster, H. A., Dering, J. C.: J. Appl. Phys. *39*, 4486 (1968)
58) For example, Kowalik, J., Osborne, M. R.: Methods for unconstrained optimization prob-lems. New York: American Elsevier Publ. 1968
59) Spendley, W., Hext, G. R., Himsworth, F. R.: Technometrics *4*, 441 (1962)
60) Nelder, J. A., Mead, R.: Computer J. *7*, 308 (1965)
61) Ogawa, K., Horii, F., Hyon, S.-H., Kitamaru, R., Yasuda, T., Okuno, T.: Polym. Prepr., Japan *26*, No. *2*, 343 (1977)
62) Kitamaru, R., Mandelkern, L.: J. Polym. Sci. Polym. Phys. Ed. *8*, 2079 (1970)
63) Mandelkern, L., Allou, Jr., A. L., Gopalan, N.: J. Phys. Chem. *72*, 309 (1968)
64) Anderson, F. R.: J. Appl. Phys. *35*, 64 (1964)
65) Mandelkern, L., Price, J. M., Gopalan, M., Fatou, J. G.: J. Polym. Sci. Part A-2, *4*, 385 (1966)
66) Kitamaru, R., Horii, F., Hyon, S.-H.: J. Polym. Sci. Polym. Phys. Ed. *15*, 821 (1977)
67) Kitamaru, R., Horii, F., Hyon, S.-H.: ACS Polym. Prepr. *17* (2), 549 (1976); revised results are used in this review, considering the effect of modulation amplitude
68) Mandelkern, L., Chow, C. Y.: private communication; Chow, C. Y.: A broad-line NMR study of molecular weight fractions of bulk polyethylene. Ph. D. dissertation, Florida State University 1973
69) Mandelkern, L.: Characterization of materials in research, ceramics and polymers. Syracuse, New York: Syracuse Univ. Press 1975, chap. 13
70) Hendus, H., Schneel, G.: Kunststoffe *55*, 69 (1961)
71) Okada, T., Mandelkern, L.: J. Polym. Sci. Part A-2, *5*, 239 (1967)
72) Gopalan, M. R., Mandelkern, L.: J. Polym. Sci. Part B, *5*, 925 (1967)
73) Mandelkern, L.: J. Phys. Chem. *75*, 3909 (1968)
74) Fatou, J. G., Mandelkern, L.: J. Phys. Chem. *69*, 417 (1965)
75) Hyon, S.-H., Horii, F., Kitamaru, R.: Bull. Inst. Chem. Res. Kyoto Univ. *55*, 248 (1977)
76) Hyndman, D., Origlio, G. G.: J. Polym. Sci. *39*, 556 (1959)
77) Stein, R. S., Noriss, F. H.: J. Polym. Sci. *21*, 381 (1956)
78) Stein, R. S.: J. Polym. Sci. *31*, 327 (1958)
79) Glenz, W., Morosoff, N., Peterlin, A.: J. Polym. Sci. Polym. Lett. Ed. *9*, 211 (1971)
80) Flory, P. J.: J. Amer. Chem. Soc., *84*, 2857 (1962)

Received March 29, 1977
S. Okamura (editor)

Author Index Volumes 1–26

B. Vollmert

Polymer Chemistry

Translated from the German by E. H. Immergut
1973. 630 figures. XVII, 652 pages
ISBN 3-540-05631-9

This book gives a comprehensive coverage of the
synthesis of polymers and their reactions, structure,
and properties. The treatment of the reactions used in
the preparation of macromolecules and in their transformation into crosslinked materials is particularly
detailed and complete. The book also gives an up-to-
date presentation of other important topics, such as
enzymatic and protein synthesis, solution properties
of macromolecules, polymer crystallization, and
properties of polymers in the solid state.
The content and presentation of Professor Vollmert's
book is more encompassing than most existing
treatises, and its numerous figures and tables convey
a wealth of data, never, however, at the expense of
intellectual clarity or educational value.
The presentation is mainly on a fundamental and
general level and yet the reader — student or profes-
sional — is gradually and almost casually introduced
to all important natural and synthetic polymers.
Complicated phenomena are explained with the aid
of the simplest available examples and models in
order to ensure complete understanding. However,
the reader is also encouraged to think for himself and
even to criticize the author's point of view.
All of the chapters have been revised and enlarged
from the German edition, and many of the sections
are entirely new.

Springer-Verlag
Berlin Heidelberg New York

B. Rånby, J. F. Rabek

ESR Spectroscopy in Polymer Research

1977. 356 figures, 29 tables. XIV, 410 pages
(Polymers/Properties and Applications, Volume 1)
ISBN 3-540-08151-8

The main purpose of this book is to collect the present
available information on the applications of electron
spin resonance (ESR) spectroscopy in polymer research.
The book has been written both for those who want
an introduction to this field, and for those who are
already familiar with ESR and are interested in
application to polymers. Therefore, the fundamental
principles of ESR spectroscopy are first outlined, the
experimental methods including computer applications
are described in more detail, and the main emphasis
is on the application of ESR methods to polymer
problems. The authors hope that this book will provide a useful source of information by giving a
coherent treatment and extensive references to
original papers, reviews, and discussions in mono-
graphs and books. In this way we hope to encourage
polymer chemists, organic chemists, biochemists,
physicists, and material scientists to apply ESR
methods to their research problems. (2519 references).

Springer-Verlag
Berlin Heidelberg New York